Joel Dorman Steele

Popular physics

Joel Dorman Steele

Popular physics

ISBN/EAN: 9783743318120

Manufactured in Europe, USA, Canada, Australia, Japa

Cover: Foto ©berggeist007 / pixelio.de

Manufactured and distributed by brebook publishing software (www.brebook.com)

Joel Dorman Steele

Popular physics

HYSICS

BY

DORMAN STEELE, Ph.D., F.G.S.

AUTHOR OF FOURTEEN-WEEKS SERIES IN NATURAL SCIENCE

The works of God are fair for naught,
Unless our eyes, in seeing,
See hidden in the thing the thought
That animates its being.

A.B.C

NEW YORK ∴ CINCINNATI ∴ CHICAGO
AMERICAN BOOK COMPANY

Steele's Science Series.

Hygienic Physiology,	$1.00
Hygienic Physiology, abridged,	.50
New Descriptive Astronomy,	1.00
Popular Physics,	1.00
Popular Chemistry,	1.00
Popular Zoology,	1.20
Fourteen Weeks in Botany,	1.00
Fourteen Weeks in Chemistry,	1.00
Fourteen Weeks in Geology,	1.00
Fourteen Weeks in Physics,	1.00
Fourteen Weeks in Philosophy (old),	1.00
Fourteen Weeks in Physiology,	1.00
Fourteen Weeks in Zoology,	1.00
Manual of Science (New Key),	1.00

Copies sent, postpaid, on receipt of price.

Copyright, 1888, by A. S. BARNES & CO.

Pop. Phys.

Printed by
A. S. Barnes & Company
New York, U. S. A.

AUTHOR'S PREFACE.

THIS work has grown up in the class-room. It contains those definitions, illustrations, and applications which seemed at the time to interest and instruct the author's pupils. Whenever any explanations fixed the attention of the learner, it was laid aside for future use. Thus, by steady accretions, the process has gone on until a book is the result.

As Physics is generally the first branch of Natural Science pursued in schools, it is important that the beginner should not be wearied by the abstractions of the subject, and so lose interest in it at the very start. The author has therefore endeavored to use such simple language and practical illustrations as will attract the learner, while he is at once led out into real life. From the multitude of philosophical principles, only those have been selected which are essential to the information of every well-read person. Within the limits of a small text-book, no subject can be exhaustively treated. This is, however, of less importance now, when every teacher feels that he must of necessity be above and beyond any school-work in the fullness of his information. The object of an elementary work is not to advance the peculiar ideas of any person, but simply to state the currently-accepted facts and theories. The time-honored classifications recognized in all scientific works, have been retained. In order to familiarize the pupil with the metric system, now generally used by scientific men, it is continually employed in the problems. The notes contain many illustrations and additional suggestions, but their great value will appear in the descriptions of simple experiments which are within the reach of any pupil.

New plates being required for this edition, the author has taken the opportunity thoroughly to revise the entire work. By carefully comparing the criticisms of teachers, he has tried to obtain the "parallax" of all its statements and methods, and to eliminate, as far as possible, the errors growing out of his "personal equation." Hearty thanks are tendered to the many friends of the book who, by their suggestions and criticisms, have so greatly added to the value of this revision. To name them all in this Preface would be impossible, and to discriminate would be invidious. The author can not, however, allow the opportunity to pass without expressing his profound sense of obligation. By untiring study and the continued help of his friends, he hopes thus, year by year, to make the series more and more worthy the favor which his fellow-teachers have so abundantly bestowed upon it. Happy indeed will he be if he succeed in leading some young mind to become a lover and an interpreter of Nature, and thus come at last to see that Nature herself is but a "thought of God."

PUBLISHERS' PREFACE.

THE series of elementary text-books in science, written by the late Dr. J. Dorman Steele, attained an extraordinary degree of popularity, due to the author's attractive style, his great skill in the selection of material suited to the demands of the schools for which the books were intended, his sympathetic spirit toward both teachers and pupils, and his earnest Christian character, which was exhibited in all his writing.

Shortly before his death, finding his health too feeble to permit of extra labor, the author requested Dr. W. Le C. Stevens, Professor of Physics in the Packer Collegiate Institute, Brooklyn, to revise the text-book in Physics, as important advances in this department of science had been made since the issue of the edition of 1878. In performing this work, Professor Stevens has endeavored to impose the least possible modification upon the peculiar style of the author. Nevertheless, every chapter has received some alterations and slight enlargement. In a work intended for higher classes, the reviser would naturally make the treatment of every subject more thorough; but this would unfit the present book for the schools to which it was originally adapted. It is difficult, moreover, to combine a strictly popular style with that precision which is demanded by advanced students in exact science. But although the field in a rudimentary text-book is limited, it is thought that there are no important errors of statement in the present hand-book.

In order to distinguish this revision from the older editions, the name is changed to "Steele's Popular Physics."

Publishers' Notice.

The publishers of this book will still issue the former edition, known as "Steele's Fourteen Weeks in Physics," for classes already organized and for teachers who may prefer to continue its use. Any book-seller can obtain the former edition if requested to do so.

PUBLISHERS' PREFACE.

THE series of elementary text-books in science, written by the late Dr. J. Dorman Steele, attained an extraordinary degree of popularity, due to the author's attractive style, his great skill in the selection of material suited to the demands of the schools for which the books were intended, his sympathetic spirit toward both teachers and pupils, and his earnest Christian character, which was exhibited in all his writing.

Shortly before his death, finding his health too feeble to permit of extra labor, the author requested Dr. W. Le C. Stevens, Professor of Physics in the Packer Collegiate Institute, Brooklyn, to revise the text-book in Physics, as important advances in this department of science had been made since the issue of the edition of 1878. In performing this work, Professor Stevens has endeavored to impose the least possible modification upon the peculiar style of the author. Nevertheless, every chapter has received some alterations and slight enlargement. In a work intended for higher classes, the reviser would naturally make the treatment of every subject more thorough; but this would unfit the present book for the schools to which it was originally adapted. It is difficult, moreover, to combine a strictly popular style with that precision which is demanded by advanced students in exact science. But although the field in a rudimentary text-book is limited, it is thought that there are no important errors of statement in the present hand-book.

In order to distinguish this revision from the older editions, the name is changed to "Steele's Popular Physics."

Publishers' Notice.

The publishers of this book will still issue the former edition, known as "Steele's Fourteen Weeks in Physics," for classes already organized and for teachers who may prefer to continue its use. Any book-seller can obtain the former edition if requested to do so.

SUGGESTIONS TO TEACHERS.

STUDENTS are expected to obtain information from this book, without the aid of questions, as they must always do in their general reading. When the subject of a paragraph is announced, the pupil should be prepared to tell all he knows about it. He should *never be allowed to answer a question*, except it be a short definition, *in the language of the book*. The diagrams and illustrations, as far as possible, should be drawn upon the blackboard and explained. Although pupils may, at first, manifest an unwillingness to do this, yet in a little time it will become an interesting feature of the recitation.. In his own classes, the author has been accustomed to *place upon the blackboard the analysis of each chapter of the book, and require the pupils to recite from that*, without the interposition of questions, except such as were necessary to bring out the topic more clearly, or to throw a side light upon it. Where the analysis given in the book does not include all the minor points of the lesson, the pupils can easily supply the omission. The "Practical Questions" given at the close of each general subject, have been found a profitable exercise in awakening inquiry and stimulating thought. They may be used at the pleasure of the instructor. The equations contained in the text are designed to be employed in the solution of the problems.

It should constantly be borne in mind that, as far as possible, every question and principle should be submitted to Nature for a direct answer by means of an experiment. Pupils should be encouraged to try the simple illustrations necessary. The student who brings in a bit of apparatus made by himself, does better than if he were merely to memorize pages of text.

SUGGESTIONS TO TEACHERS.

The following works, to which the author acknowledges his obligation for valuable material, will be useful to teacher as well as pupil, in furnishing additional illustrations and in elucidating difficult subjects, viz.: Tait's "Recent Advances on Physical Science"; Arnott's "Elements of Physics" (7th ed.); Stewart's "Elementary Physics," also his "Conservation of Energy," and "Treatise on Heat"; Atkinson's "Deschanel's Natural Philosophy"; Lockyer's "Guillemin's Forces of Nature"; Herschel's "Introduction to the Study of Physical Science"; Tomlinson's "Introduction to the Study of Natural Philosophy"; Beale's "How to Work with the Microscope"; Schellen's "Spectrum Analysis"; Roscoe on "Spectrum Analysis"; Lockyer's "The Spectroscope," and "Studies in Spectrum Analysis"; Airy's "Geometrical Optics"; Nugent's "Optics"; "Chevreul on Colors"; Thomson and Tait's "Natural Philosophy"; Maxwell's "Electricity and Magnetism"; Silvanus Thompson's "Lessons in Electricity and Magnetism"; Faraday's "Forces of Matter"; Youmans' "Correlation of Physical Forces"; Maury's "Physical Geography of the Sea"; Atkinson's "Ganot's Physics"; Silliman's "Physics"; Tyndall's Lectures on Light, Heat, Sound, Electricity, also his "Forms of Water"; Snell's "Olmsted's Philosophy" (revised edition); Loomis' "Meteorology"; Miller's "Chemical Physics"; Urbanitzky's "Electricity in the Service of Man"; Cooke's "Religion and Chemistry"; Daniell's "Principles of Physics"; Anthony and Brackett's "Text-book of Physics," and also numerous works named in the "Reading References" at the close of each general division. They may be procured of the publishers of this book. The pupil should continually be impressed with the thought that the text-book only introduces him to a subject, which he should seek every opportunity to pursue in larger works and in treatises on special topics.

The editor will be pleased to correspond with teachers concerning the apparatus for the performance of the experiments, or with reference to any of the "Practical Questions."

TABLE OF CONTENTS.

	PAGE
I.—INTRODUCTION	1
I.—General Definitions	3
II.—General Properties of Matter	6
III.—Specific Properties of Matter	10
II.—MOTION AND FORCE	19
III.—ATTRACTION	41
I.—Molecular Forces	43
II.—Gravitation	55
IV.—ELEMENTS OF MACHINES	79
V.—PRESSURE OF LIQUIDS AND GASES	99
I.—Hydrostatics	101
II.—Hydrodynamics	121
III.—Pneumatics	129
VI.—SOUND	151
VII.—LIGHT	189
VIII.—HEAT	241

CONTENTS.

	PAGE
IX.—MAGNETISM	279
X.—ELECTRICITY	291
XI.—APPENDIX	355
1.—QUESTIONS	357
2.—INDEX	375

I.

INTRODUCTION.

"We have no reason to believe that the sheep or the dog, or indeed any of the lower animals, feel an interest in the laws by which natural phenomena are regulated. A herd may be terrified by a thunder-storm; birds may go to roost, and cattle return to their stalls during a solar eclipse; but neither birds nor cattle, so far as we know, ever think of inquiring into the causes of these things. It is otherwise with man. The presence of natural objects, the occurrence of natural events, the varied appearances of the universe in which he dwells, penetrate beyond his organs of sense, and appeal to an inner power of which the senses are the mere instruments and excitants. No fact is to him either final or original. He can not limit himself to the contemplation of it alone, but endeavors to ascertain its position in a series to which the constitution of his mind assures him it must belong. He regards all that he witnesses in the present as the afflux and sequence of something that has gone before, and as the source of a system of events which is to follow. The notion of spontaneity, by which in his ruder state he accounted for natural events, is abandoned; the idea that nature is an aggregate of independent parts also disappears, as the connection and mutual dependence of physical powers become more and more manifest; until he is finally led to regard Nature as an organic whole, as a body each of whose members sympathizes with the rest, changing, it is true, from age to age, but without any real break of continuity, or interruption of the fixed relations of cause and effect."

Tyndall.

ANALYSIS OF THE INTRODUCTION.

INTRODUCTION.

- **I. GENERAL DEFINITIONS.**
  1. Of Matter, Body, and Substance.
  2. General and Specific Properties.
  3. The Atomic Theory.
  4. Physical and Chemical Changes.
  5. Energy.
  6. Physical and Chemical Forces.
  7. Definition of Physics and Chemistry.

- **II. GENERAL PROPERTIES OF MATTER.**
  1. Extension and its Measurement.
  2. Impenetrability.
  3. Divisibility.
  4. Porosity.
  5. Indestructibility.

- **III. SPECIFIC PROPERTIES OF MATTER.**
  1. Ductility.
  2. Malleability.
  3. Tenacity.
  4. Elasticity.
     1. *Compression.*
     2. *Expansion.*
     3. *Torsion.*
     4. *Flexure.*
  5. Hardness.
  6. Brittleness.

INTRODUCTION.

I. GENERAL DEFINITIONS.

1. Matter.—Whatever occupies space is called *matter*. A definite portion of matter is termed a *body*.—*Examples:* A lake, a dew-drop, a quart of oil, an anvil, a pendulum. A particular kind of matter is styled a *substance*.—*Examples:* Gold, wood, stone, oxygen.

2. General and Specific Properties.—A *general property* of matter is a quality that belongs to all substances.—*Example:* Extension. A *specific property* is one which distinguishes particular substances.—*Examples:* The yellow color of gold, the brittleness of glass, the sweetness of sugar. These properties are so distinctive that we say, "yellow as gold," "brittle as glass," "sweet as sugar."

3. The Atomic Theory supposes

(1.) That the smallest particle of matter we can see is composed of still smaller particles or *molecules* (tiny masses),* each possessing the specific properties of the substance to which it belongs.

---

* A molecule is a group of atoms held together by chemical force, and is the smallest particle of a substance which can exist by itself. Even in a simple substance, *i. e.*, one in which the atoms are all of one kind, it is thought that they are generally clustered in molecules. (See "Chemistry," p. 4.) In water, the molecules are the small masses which, when driven

(2.) That each molecule consists of still smaller portions, called *atoms*,* which are regarded as indivisible and unchangeable.—*Examples*: A molecule of water is made up of two atoms of hydrogen and one of oxygen. A molecule of salt consists of one atom of chlorine and one of sodium. The smallest piece of salt contains many molecules. By dissolving in water, we divide it into its separate molecules, and the solution has a briny taste, because each one possesses the savor of salt.

4. Physical and Chemical Changes.—A *physical change* is one that does not affect the composition of the molecule, and so does not alter the specific properties of a substance.—*Examples*: The falling of a stone, the dissolving of sugar in water. A *chemical change* is one that implies the re-arrangement of the atoms into new molecules and so destroys the specific properties of a substance.—*Examples*: The rusting of iron, the burning of coal.

5. Energy.—The power of producing change of any kind is called *energy*. Heat, light, electricity, etc.,

---

apart, form steam. In a gas, they move about with great velocity, colliding with one another and with the sides of the containing vessel. The collisions of the molecules with the walls of the vessel account for the pressure which the gas exerts upon them.

* Animalcules furnish a striking illustration of the minuteness of atoms. In the drop of stagnant water that clings to the point of a needle, swarming legions swim as in an ocean, full of life, frisking, preying upon one another, waging war, and re-enacting the scenes of the great world about them. These tiny animals possess organs of digestion and assimilation. Their food, coursing in channels more minute than we can conceive, may be composed of solid as well as liquid matter; and finally, at the lowest extreme of this descending series, we come to the atoms of which the matter itself is composed. The most powerful of microscopes fails completely to reveal the separate molecule.

are different forms of energy. Each is capable of producing change in matter.—*Examples:* Heat melts ice; light blackens silver chloride; a wire becomes hot when an electric current passes through it. The same energy may be manifested successively in different forms.—*Examples:* Heat is converted by the steam-engine into motion; this motion imparted to a dynamo is converted into electricity, which may be further transformed into light, heat and magnetism.

6. Physical and Chemical Forces.—When energy is manifested as an attraction or repulsion we speak of it as a *force*, and we distinguish between physical and chemical forces as the changes produced by them are physical or chemical.—*Examples:* The attraction of an electrified body for small bits of paper is a physical force, and so also the attraction of a magnet for one end of a compass-needle, or the repulsion for the other end. The attraction of iron for oxygen in the formation of iron-rust is a chemical force.

7. Physics and Chemistry.—The former treats of physical changes, the latter of chemical changes in matter. The unit of the physicist is the molecule, of the chemist the atom.

PRACTICAL QUESTIONS.

1. Name some specific property of coal; ink; chalk; grass; tobacco; snow.
2. My knife-blade is magnetized, so that it will pick up a needle; is that a physical or a chemical change?
3. Is it treated in Physics or Chemistry?
4. Is the burning of coal a physical or a chemical change?
5. The production of steam? The formation of dew?

6. The falling of a stone? The growth of a tree?
7. The flying of a kite? The chopping of wood?
8. The explosion of powder? The boiling of water?
9. The melting of iron? The drying of clothes?
10. The freezing of water? The dissolving of sugar?
11. The forging of a nail? The making of bread?
12. The sprouting of a seed? The decay of vegetables?
13. The condensation of steam?

II. GENERAL PROPERTIES OF MATTER.

THERE are two essential properties without which matter is inconceivable. These are extension and impenetrability.

1. Extension is the property of occupying space. The amount of space a body occupies is called its *volume*.

Measurement of Extension.—A body has three dimensions: length, breadth, and thickness. To measure these, some standard is required. The standard of length popularly in use in England and the United States is the yard. Its length is the distance between two lines on a certain bar of bronze, kept in London and measured at a certain temperature, 62° F. (see p. 248). There is only one yard in the world; all that we call yards are imperfect copies from it. The yard is inconveniently divided into three feet, or thirty-six inches. The standard of length used in France, and by scientific men throughout the world, is the meter.* Its length is nearly, but not exactly,

---

* The meter is divided into ten decimeters (*dm.*); each of these into ten centimeters (*cm.*); and each of these into ten millimeters (*mm.*). In Fig. 1 is

GENERAL PROPERTIES OF MATTER. 7

$\frac{1}{40\,000\,000}$ of an entire meridian of the earth. There is only one meter in the world. It is the length of a certain bar of platinum, kept in Paris, and measured at the temperature of melting ice. Most copies of the meter and yard are accurate enough for the purposes to which they are applied.

2. Impenetrability is the property of so occupying space as to exclude all other matter.* No two bodies can occupy the same space at the same time. A book lies upon the table before me; no human power is able to place another in the same spot, until the first book

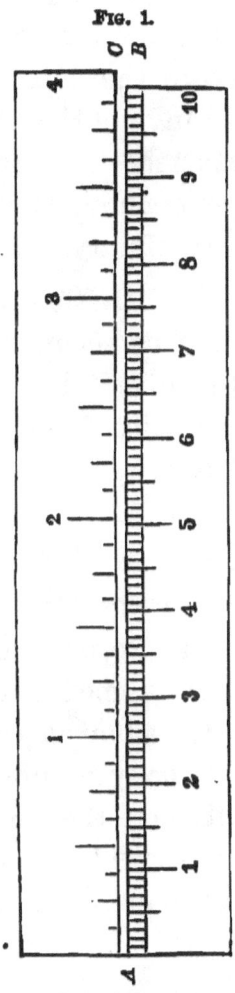

FIG. 1.

Comparison of Metric and English Measures of Length.

shown a line, *AB*, whose length is a decimeter, divided into centimeters and millimeters. At the side of it is another line, *AC*, slightly longer. It is made up of four inches, divided into halves, quarters, and eighths. The length of the meter is about 39.37 inches, or nearly 1.1 yard.

For the measurement of surface, we use square meters (*sq. m.*), square centimeters (*sq. cm.*), etc.

The unit adopted for the measurement of volume is the cubic decimeter. It is called a *liter*. A vessel that contains just a liter of water will hold a little more than a quart of the same liquid. Since the liter has a length, breadth, and thickness of one decimeter, it contains 1,000 cubic centimeters.

* In common language, we say a needle penetrates cloth, a nail enters wood, etc.; but a moment's examination shows that they merely push aside the fibers of the cloth or wood, and so press them closer together. With care we can drop a quarter of a pound of shingle-nails into a tumbler brimful of water, without causing it to overflow. The surface of the water, however, becomes convex.

is removed. I attempt to fill a bottle through a closely-fitting funnel; but before the liquid can run in, the air must gurgle out, or the water will trickle down the outside of the bottle.

In addition to these two essential properties of matter, there are others which have been found to be general, such as divisibility, porosity, and indestructibility.

3. Divisibility is that property by which a body may be separated into parts. The extent to which the divisibility of matter may be carried is almost incredible.*—*Example:* A grain of strychnine will flavor 1,750,000 grains of water; hence there will be in each grain of the liquid only $\frac{1}{1750000}$ of a grain of strychnine, yet this amount can be distinctly tasted.

4. Porosity is the property of having pores. By this is meant not the *sensible* pores to which we refer when in common language we speak of a porous body, as bread, wood, unglazed pottery, a sponge, etc., but the finer or *physical* pores. The latter are as invisible to the eye as the atoms themselves, and are caused by the fact that the molecules of which a body is composed are not in actual contact, but

---

* Newton estimated that the film of a soap-bubble at the instant of breaking is less than $\frac{1}{1750000}$ of an inch thick. Pure water will acquire the requisite viscidity for making bubbles by adding only $\frac{1}{100}$ part of soap. It is evident that there must be at least one molecule of soap in every cubic $\frac{1}{1750000}$ of an inch of the film, and that the molecule must be smaller than one hundredth of a cubic $\frac{1}{1750000}$ of an inch, *i. e.*, than $\frac{1}{175000000}$ of a cubic inch. Now a molecule of soft-soap (if it is a pure potassium stearate, "Chemistry," p. 207) contains 56 atoms, and this point must be reached before we come to the possible limit of divisibility.

GENERAL PROPERTIES OF MATTER. 9

are separated by minute spaces.*—*Examples:* To a bowl-full of water it is easy to add a quantity of fine salt without the liquid running over. Only care must be taken to drop in the salt slowly, giving time for it to dissolve and the bubbles of air to pass off. When the water has dissolved all the salt it can, we can still add other soluble solids.†—In testing large cannon by hydrostatic pressure (p. 104), water is forced into the gun until it oozes through the thick metal and covers the outside of the gun like froth, then gathers in drops and runs to the ground in minute streams.‡

It is in virtue of these physical pores that a body changes in volume when warmed or cooled. The molecules become farther apart or nearer as heat is

---

* These spaces are so small that they can not be discerned with the most powerful microscope, yet it is thought that they may be very large when compared with the size of the atoms themselves. If we imagine a being small enough to live on one of the atoms near the center of a stone, as we live on the earth, then we are to suppose that he might possibly see the nearest atoms at great distances from him, as we see the moon and stars, and might perchance have need of a fairy telescope to examine them, as we investigate the heavenly bodies. It is impossible, however, for us to have any definite knowledge on such a topic.

† In this case we suppose that the particles of salt are smaller than those of water, and those of the different substances used are smaller than those of salt. The particles of salt fill the spaces between the particles of water, and the others occupy the still smaller spaces left between the particles of salt. We may better understand this if we suppose a bowl filled with oranges. It will hold a quantity of peas, then of gravel, then of fine sand, and lastly some water.

‡ In the course of some experiments performed during the eighteenth century at the Florence Academy, Italy, hollow globes of silver were filled with water and placed in a screw-press. The spheres being flattened, their size was diminished, and the water oozed through the pores of the metal. The philosophers of the day thought this to show that water is incompressible. We now see that it proved only that silver has pores larger than the molecules of water.

applied or withdrawn. We can not conceive this to be possible if they are in perfect contact.

5. Indestructibility is the property which renders matter incapable of being destroyed. We can not conceive of the annihilation of matter. We may change its form, but we can not deprive it of existence.—*Example:* We cut down a tree, saw it into boards, and build a house. The house burns, and only little heaps of ashes remain. Yet in the ashes, and in the smoke of the burning building, exist the identical atoms, which have passed through these various forms unchanged.*

III. SPECIFIC PROPERTIES OF MATTER.

AMONG the most important specific properties of matter are ductility, malleability, tenacity, elasticity, hardness, and brittleness.

1. Ductility.—A ductile body is one which can be drawn into wire. Fig. 2 represents a machine for making wire. *B* is a steel drawing-plate pierced with a series of gradually diminishing holes. A rod of iron, *A*, is hammered at the end so as to pass

---

* Walter Raleigh, while smoking in the presence of Queen Elizabeth, offered to bet her majesty that he could tell the weight of the smoke that curled upward from his pipe. The wager was accepted. Raleigh quietly finished, and then weighing the ashes, subtracted this amount from the weight of the tobacco he had placed in the pipe, thus finding the weight of the smoke. When we reach the subject of combustion in chemistry, we shall be able to detect Raleigh's mistake. The smoke and the ashes really weighed more than the original tobacco, since the oxygen of the air had combined with the tobacco in burning.

through the largest. It is then grasped by a pair of pincers, *C*, and, by turning the crank, *D*, is drawn through the plate, diminished in diameter and proportionately increased in length. The tenacity of the metal is greatly improved by the

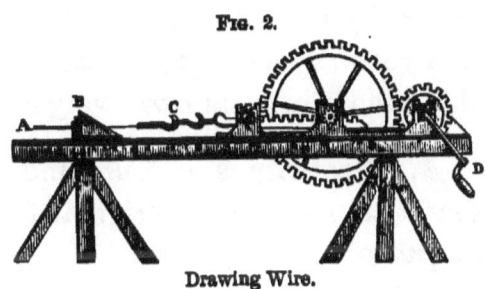

Fig. 2.

Drawing Wire.

process of drawing, so that a cable of fine wire is stronger than a chain or bar of the same weight. Gold, silver, and platinum are the most ductile metals. A silver rod an inch thick, covered with gold leaf, may be drawn to the fineness of a hair and yet retain a perfect coating of gold, three ounces of the latter metal making 100 miles of the gilt-thread used in embroidery. Platinum wire has been drawn so fine that, though it is nearly three times as heavy as iron, a mile's length weighed only a grain.

2. Malleability.—A malleable body is one which can be hammered or rolled into sheets.—*Example:* Gold may be beaten until it is only $\frac{1}{300000}$ of an inch thick. It would require 1,800 such leaves to equal the thickness of common printing-paper.* Copper is

---

* An ingot of gold is passed many times between steel rollers, which are so adjusted as to be continually brought nearer together. The metal is thus reduced to a ribbon about $\frac{1}{10}$ of an inch thick. This is cut into inch squares, 150 of which are piled up alternately with leaves of strong paper four inches square. A workman with a hammer beats the pile until the gold is spread to the size of the leaves. Each piece is next quartered, and the 600 squares are placed between leaves of goldbeaters'

so malleable, that a workman can hammer out a kettle from a solid block.

3. Tenacity.—A tenacious body is one which can not easily be pulled apart. Iron possesses this quality in a remarkable degree. Steel wire will sustain many thousand times its own weight.

4. Elasticity is of four kinds, according as a body tends to resume its original form when *compressed, extended, twisted,* or *bent*.

(1.) ELASTICITY OF COMPRESSION.—Many *solids*, as iron, glass, and caoutchouc, are highly elastic.—*Example:* Spread a thin coat of oil on a smooth marble slab. If an ivory ball be dropped upon it, the size of the impression will vary with the distance at which the ball is held above the table. This shows that the ivory is flattened, somewhat like a soap-bubble when it strikes a smooth surface and rebounds.

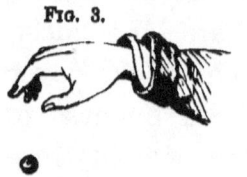
Fig. 3.

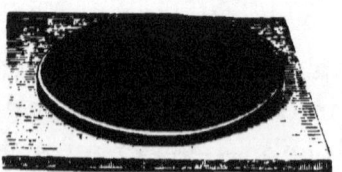
Elasticity of Ivory.

Liquids are compressed with great difficulty, so that for a long time they were considered incompressible. When the force is

skin and pounded. They are then taken out, spread by the breath, cut, and the 2,400 squares pounded as before. They are finally trimmed and placed between tissue-paper in little books, each of which contains twenty-five gold leaves.

removed, they regain their exact volume, and are therefore perfectly elastic.

Gases are easily compressed, and are also perfectly elastic. A pressure of 15 lbs. to the square inch reduces the volume of water only $\frac{1}{20400}$, whereas it diminishes the volume of a perfect gas $\frac{1}{4}$. A gas may be kept compressed for years, but on being released will instantly return to its original form.

(2.) ELASTICITY OF EXPANSION is possessed largely by very many substances.—*Examples:* India rubber, when stretched, tends to fly back to its original dimensions. A drop of water hanging to the nozzle of a bottle may be touched by a piece of glass and drawn out to considerable length, but when let go it will resume its spherical form.

(3.) ELASTICITY OF TORSION is the tendency of a thread or wire which has been twisted, to untwist again. If a weight be suspended by means of a steel wire, twisted around and then released, it constitutes a torsion pendulum. (Fig. 4.)

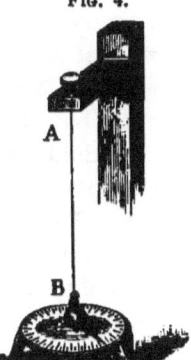

FIG. 4.

The Torsion Pendulum.

(4.) ELASTICITY OF FLEXURE is the property ordinarily meant by the term elastic. Many solids possess this quality, within certain limits, to a high degree. Swords have been made which could be bent into a circle without breaking. Watch-springs, bows, cushions, etc., are useful because of their elasticity. Glass, though brittle, is one of the most elastic substances known.

5. Hardness.—One body is harder than another when it will scratch or indent it. This property does not depend on density.*—*Examples:* Gold is about 2¼ times as dense as iron, yet it is much softer.—Mercury is a liquid, yet it is almost twice as dense as steel.—The diamond is the hardest known substance, yet it is not one third as heavy as lead.

6. Brittleness.—A brittle body is one that is easily broken. This property is a frequent characteristic of hard bodies.—*Example:* Glass will scratch pure iron, yet it is extremely brittle.

SUMMARY.

MATTER is that which occupies space. A separate portion is called a body, and a particular kind a substance. A general property of matter belongs to all substances, and a specific one to particular kinds. Matter is composed of very minute atoms. A group of atoms forms a molecule, in which reside the specific properties of a substance. A physical change never affects the molecule, but a chemical change breaks it up, and so makes new combinations possible. Physics deals with physical forces and changes; Chemistry with chemical attraction, and chemical changes. Extension and impenetrability are the essential properties of matter. Extension is the property of occupying space. The amount of space a body fills is its volume. In virtue of impenetrability two bodies can not occupy the same space at the same time. The divisibility of matter is without perceptible limit so far as we know. Porosity is the property in virtue of which the molecules of a body are not in absolute contact. Indestructibility prohibits the extinction of matter by man. Duc-

---

* A *dense* body has its molecules closely compacted. The word *rare*, the opposite of dense, is applied to gases. *Mass*, or the quantity of matter a body contains, should be distinguished from weight or size.

tility, malleability, tenacity, elasticity, hardness, and brittleness are the principal specific properties of matter. A ductile body can be drawn into wire; gold, silver, and platinum are the most noted for this property. A malleable body can be hammered into sheets; gold possesses this quality in a remarkable degree. A tenacious body resists pulling apart; iron is the best example. An elastic body permits a play of its particles, so that they return to their original position when the disturbing force is removed. A hard body can not easily be indented. A brittle body is readily broken.

HISTORICAL SKETCH.

IN ancient times, any seeker after truth was termed a philosopher (a lover of wisdom), and philosophy included all investigations concerning both mind and matter. In the fourth century B.C., Plato assumed that there are two principles, matter and form, which by combining produce the five elements, earth, air, fire, water, and ether. Aristotle, his pupil, established the first philosophical ideas concerning matter and space. But the method of study generally pursued for 2,000 years was one of pure metaphysical speculation. Observation had no place, but the philosophers made up a theory, and then accommodated facts to it. They guessed about the real essence of things, as to whether matter exists except when perceived by the mind,* and how a change in matter can produce a change in mind. In 1620, Bacon published his "Novum Organum," advocating the inductive method of studying nature. He argued that the philosopher should seek to benefit mankind, and that, instead of wasting his time in sterile and ingenious theories about the world and matter, he should watch the phenomena of life, gather facts, and then reasoning from effects back to their causes, reach the general law. This work

---

* Dr. Johnson once remarked to a gentleman who had been defending the theory that there is no external world, as he was going away, "Pray, sir, don't leave us, for we may perhaps forget to think of you, and then you will cease to exist."

is commonly said to have established the modern method of investigation. Ptolemy, Archimedes, Galileo, and other physicists, however, had long before proved its value.

The Atomic Theory was propounded by Democritus, in the fifth century B.C., and twenty-two centuries later elaborated by Dalton, an English physicist. The grander generalization and development of this law was advanced in 1811 by Avogadro, an Italian, and afterward extended by the French philosopher, Ampere. The latter asserted that "equal volumes of all substances, when in the gaseous form and under like conditions, contain the same number of molecules." For half a century this hypothesis was ignored. Its adoption within the past thirty years produced fundamental changes in chemistry. Displacing the old it gave birth to a new and consistent notation, and, raised to the rank of a law, it served as a criterion by which the correctness of its formulæ might be tested.

The history of the establishment of a standard of measures is a curious one. Anciently, length was referred to some portion of the human body, as the foot; the cubit (the length of the fore-arm from the elbow to the end of the middle finger); the finger's length or breadth; the hand's breadth; the span, etc. In England, Henry I. (1120) ordered that the ell, the ancient yard, should be the exact length of his arm. Afterward a standard yard-stick was kept at the Exchequer in London; but it was so inaccurate, that a commissioner, who examined it in 1742, wrote: "A kitchen poker filed at both ends would make as good a standard. It has been broken, and then repaired so clumsily that the joint is nearly as loose as a pair of tongs." In 1760, Mr. Bird carefully prepared a copy of this for the use of the government. It was not legally adopted until 1824, when it was ordered that if destroyed, it should be restored by a comparison with the length of a pendulum vibrating seconds at the latitude of London. At the great fire in London, 1834, the Parliament House was burned, and with it Bird's yard-stick. Repeated attempts were then made to find the length of the lost standard by means of the pendulum. This was found impracticable, on account of errors in the original directions. At last the British government adopted a standard prepared from the most reliable copies of Bird's yard-stick. A

HISTORICAL SKETCH. 17

copy of this was taken by Troughton, an instrument-maker of London, for the use of our Coast Survey.*

The French had previously adopted for the length of the legal foot that of the royal foot of Louis XIV., as perishable a standard as Henry's arm. In 1790, the Prince de Talleyrand proposed to the Constituent Assembly of France the foundation of a system based on a single and universal standard, which might be used by all civilized nations. The selection of this was committed to five members of the Academy of Sciences, MM. Borda, Lagrange, Laplace, Monge, and Condorcet, who decided that the ten-millionth part of a quarter of the earth's meridian should be taken as the standard of length, from which the standards of surface, volume, capacity, and weight should be derived. A trigonometric measurement was made of the arc of a meridian extending through France from Dunkirk to Barcelona, a work which occupied seven years. In 1799, an international commission was assembled at Paris, with representatives from most of the governments of Europe. They deposited at the Palace of the Archives, in Paris, the standard meter-bar of platinum, and the standard kilogram weight, made of the same metal. In English denominations the length of the meter is almost exactly 3.28 feet, or 39.37 inches; the weight of the kilogram almost exactly 2.2 pounds avoirdupois.

But after the establishment of the metric system it was found that a slight mistake had been made in the measurement of the arc of the meridian. The English, who had declined to accept the French system, discovered also a difficulty in the determination of the yard from the pendulum beating seconds. Both the yard and the meter are therefore arbitrary and not absolute standards. Copies of each have been made so carefully and distributed so widely that there is no probability of any appreciable loss resulting from the accidental destruction of the originals. The metric system is by far the simplest and best in use, but it has not yet generally supplanted other sys-

---

* A bronze bar, which has the standard length at 61.79° F., has been presented by the English government to that of the United States. According to Act of Congress, sets of weights and measures have been distributed to the governors of the several States. Both the yard and the meter are legal standards in the United States and Great Britain.

tems that retain their popularity, not on account of merit, but only because of human conservatism and the inconveniences resulting from change.

Consult Cooke's "New Chemistry," chapter on Molecules, etc.; Powell's "History of Natural Philosophy"; Buckley's "History of Natural Science"; Whewell's "History of the Inductive Sciences"; Roscoe's "John Dalton and his Atomic Theory," in Manchester Science Lectures, '73–4; "Appleton's Cyclopedia," Art. Molecules; Outerbridge's "Divisibility of Gold and Other Metals," in Popular Science Monthly, Vol. XI., p. 74; Crookes' "The Radiometer—a fresh evidence of a Molecular Universe," Popular Science Monthly, Vol. XIII., p. 1; Tait's "Recent Advances in Physical Science," Chap. XII., The Structure of Matter; Hoefer, "Histoire de la Physique et de la Chimie"; Draper's "History of Intellectual Development of Europe"; Barnard on "The Metric System."

II.

MOTION AND FORCE.

REST is nowhere. The winds that come and go, the ocean that uneasily throbs along the shore, the earth that revolves about the sun, the light that darts through space—all tell of a universal law of Nature. The solidest body hides within it inconceivable velocities. Even the molecules of granite and iron have their orbits as do the stars, and move as ceaselessly.

No energy is ever lost. It changes its form, but the eye of philosophy detects it and enables us to drive it from its hiding-place undiminished. It assumes Protean guises, but is every-where a unit. It may disappear from the earth; still—

> "Somewhere yet that atom's force
> Moves the light-poised universe."

ANALYSIS OF MOTION AND FORCE.

MOTION AND FORCE.
1. Definitions.
2. Communication of Motion.
3. Resistances to Motion.
4. First Law of Motion.
   - (1.) Inertia.
   - (2.) Momentum.
5. Second Law of Motion.
6. Third Law of Motion.
7. Composition of Motions.
8. Composition of Forces.
9. Triangle of Forces.
10. Polygon of Forces.
11. Resolution of Forces.
12. Motion in a Curve.
13. Circular Motion.
14. Reflected Motion.
15. Energy.
16. Kinetic and Potential Energy.
17. Conservation of Energy.

MOTION AND FORCE.

1. Motion is change of place. All motion, as well as rest, with which we are acquainted, is relative.—*Examples:* When we ride in the cars, we judge of our motion by the objects around us.—A man on a steamer may be in motion with regard to the shore, but at rest with reference to the objects on the deck of the vessel. *Force* is that which produces or tends to produce or to destroy motion. *Velocity* is the rate at which a body moves. When the rate is constant, *i.e.*, when the moving body passes over equal spaces in equal times, the body moves with *constant* velocity. It is expressed by the number of units of space through which the body moves in a unit of time. —*Example:* Ten miles an hour or fifteen feet a second. When the rate is variable, *i.e.*, when the spaces passed over by the body in equal times are *unequal*, the body moves with *variable* velocity.—*Example:* The velocity of a train between two stations is variable. In cases of this kind it is convenient to speak of the *average velocity* of the train, which is the ratio of the whole distance passed over to the time required for such passage. The train will evidently arrive at the second station in the same time, whether it move uniformly with the average velocity or with its actually varying velocity.

2. The Communication of Motion is not instantaneous. If I strike one end of a rail a mile long the tremor communicated from particle to particle will take a definite time to reach the other end. A stone thrown against a pane of glass shatters it, but a bullet fired through it will make only a round hole. The bullet is gone before the motion has time to be communicated to the surrounding particles. A fraction of time is required for a ball to receive the force of the exploding powder and get under full headway.

3. The Resistances to Motion are friction and the resistance of air and water. (1.) Friction is the resistance caused by the surface over which a body moves. It is of great value in common life. Without it, nails, screws, and strings would be useless; engines could not draw the cars; we could hold nothing in our hands; and we should everywhere walk as on glassy ice. (2.) The resistance which a body meets in passing through air or water is caused largely by the particles displaced.

LAWS OF MOTION.—There are three laws of motion, which were first distinctly formulated by Sir Isaac Newton.

4. First Law of Motion.—*A body set in motion will move forever in a straight line with constant velocity, unless acted on by some external force.* Obviously, no experiment will directly prove this law. Common experience, moreover, seems to contradict it; for everywhere on the surface of the earth bodies in motion show a tendency to stop. A little reflection will show, however, that these bodies are subject to the action of external forces tending to arrest the motion. It is supposed that could these resistances be removed a body once in motion would never stop.

Inertia.—The law just stated is often called the *law of inertia.* Matter has no inherent power of producing change upon itself. If a body be already in motion, force has to be expended in stopping it. If it be at rest, force is required to start it in motion. In either case we "overcome its inertia." The danger in jumping from a car in rapid motion lies in the fact that the body has the speed of the train, while the forward motion of the feet is checked by contact with the ground. It is necessary to jump as nearly as possible in the direction in which the train is moving, and be ready to run the instant the feet touch the ground. Those who do so can then gradually overcome the inertia of the body, and after a few yards can turn as they please.

Momentum.—To measure any force we must know what quantity of matter is moved, and what velocity it receives in a unit of time. The quantity of matter in a body is called its *mass.* It is not the same as weight, but is proportional to this (see p. 57), so that we speak of pounds of mass as well as pounds of weight. The product of mass by velocity is called *momentum.*[*] Thus if a mass of five pounds move

---

[*] A heavy body may be moving very slowly and yet have an immense momentum.—*Examples:* An iceberg, with a scarcely perceptible motion, will crush the strongest ship as if it were an egg-shell.—Soldiers have thought to stop a spent cannon-ball by putting a foot against it, but have found its momentum sufficient to break a leg.

On the other hand, a light body moving with a high velocity may have

with a velocity of twenty feet per second, it has one hundred units of momentum.

5. Second Law of Motion.—*A force acting upon a body in motion or at rest, produces the same effect whether it acts alone or with other forces.*—*Examples:* All bodies upon the earth are in constant motion with it, yet we act with the same ease that we should were the earth at rest.*—We throw a stone

an enormous momentum.—*Examples:* The air in a hurricane will tear up trees by the roots and level buildings to the ground.—Sand driven from a tube by steam is used for drilling and in stone-cutting, engraving, etc.

"In a rude age, before the invention of means for overcoming friction, the weight of bodies formed the chief obstacle to setting them in motion. It was only after some progress had been made in the art of throwing missiles, and in the use of wheel-carriages and floating vessels, that men's minds became practically impressed with the idea of mass as distinguished from weight. Accordingly, while almost all the metaphysicians who discussed the qualities of matter, assigned a prominent place to weight among the primary qualities, few or none of them perceived that the sole unalterable property of matter is its *mass*. At the revival of science this property was expressed by the phrase 'The inertia of matter'; but while the men of science understood by this term the tendency of the body to persevere in its state of motion (or rest), and considered it a measurable quantity, those philosophers who were unacquainted with science understood inertia in its literal sense as a quality—mere want of activity or laziness I therefore recommend to the student that he should impress his mind with the idea of mass by a few experiments, such as setting in motion a grindstone or a well-balanced wheel, and then endeavoring to stop it, twirling a long pole, etc., till he comes to associate a set of acts and sensations with the scientific doctrines of dynamics, and he will never afterward be in any danger of loose ideas on these subjects."—Maxwell's "Theory of Heat," p. 85

* A ball thrown up into the air with a force that would cause it to rise fifty feet, will ascend to that height whatever horizontal wind may be blowing.—While riding on a car, we throw a stone at some object at rest. The stone, having the motion of the train, strikes just as far ahead of the object as it would have gone had it remained on the train. In order to hit the mark, we should have aimed a little back of it.—The circus-rider wishes, while his horse is at full speed, to jump through a hoop suspended before him. He simply springs directly upward. Going forward by the momentum which he had acquired before he leaped from the horse, he passes through the hoop and alights upon the saddle again.—A person

directly at an object and hit it, yet, within the second, the mark has gone forward many feet.*—If a cannon-ball be thrown horizontally, it will fall as fast and strike the earth as soon as if dropped to the ground from the muzzle of the gun. In Fig. 5, D is an arm driven by a wooden spring, E, and turning

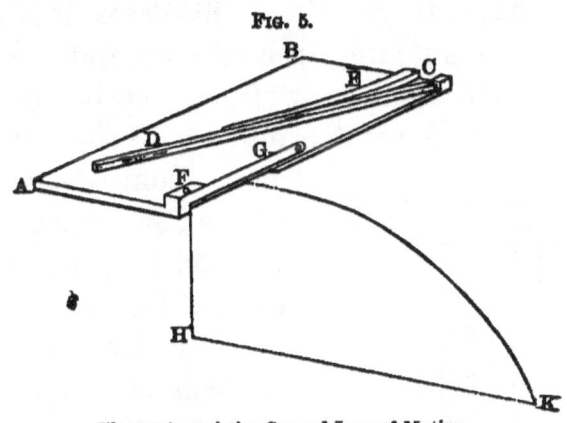

Illustration of the Second Law of Motion.

on a hinge at C. At D is a hollow containing a bullet, so placed that when the arm is sprung, the ball will be thrown in the line FK. At F is a similar ball, supported by a thin slat, G, and so arranged that the same blow which throws the ball, D, will let the ball, F, fall in the line FH. The two balls will strike the floor at the same instant.

6. Third Law of Motion.—*Action is equal to reaction, and in the contrary direction.*—*Examples:* A bird in flying beats the air downward, but the air

riding in a coach drops a cent to the floor. It apparently strikes where it would if the coach were at rest.

* The earth moves in its orbit around the sun at the rate of about eighteen miles per second. (See "Fourteen Weeks in Astronomy," p. 106.)

reacts and supports the bird.—The powder in a gun explodes with equal force in every direction, driving the gun backward and the ball forward, with the same momentum. Their velocities vary with their weights; the heavier the gun, the less will the recoil be noticed.—When we spring from a boat, unless we are cautious, the reaction will drive it from the shore.—When we jump from the ground, we tend to push the earth from us, while it reacts and pushes us from it; we separate from each other with equal momentum, and our velocity is as much greater than that of the earth as we are lighter.—We walk therefore by reason of the reaction of the ground on which we tread.

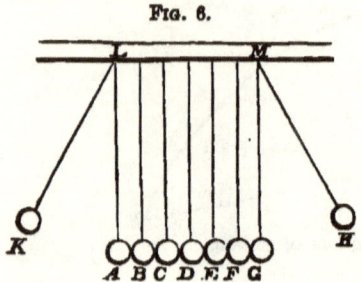

Fig. 6.

Illustration of the Third Law of Motion.

The apparatus shown in Fig. 6 consists of ivory balls hung so as to vibrate readily.* If a ball be let fall from one side, it will strike the second ball, which will react with an equal force, and stop the motion of the first, but transmit the motion to the third; this will act in the same manner, and so on through the series, each acting and reacting until the last ball is reached; this will react and then bound off, rising as high as the first ball fell (except the loss caused by resistances to motion). If two

---

* The same experiments can be performed by means of glass marbles or billiard balls placed in a groove. Better still, attach strings to glass marbles by means of mucilage and bits of paper and suspend them from a simple wooden frame.

balls be raised, two will fly off at the opposite end; if two be let fall from one side and one from the other, they will respond alternately.

7. Composition of Motions.—Let a ball at *A* (Fig. 7) be acted on by a force which would drive it in a given time to *B*, and also at the same instant by another which would drive it to *D* in the same time; the ball will move in the direction *AC*.—*Examples:* A person wishes to row a boat across a swift current which would carry him down stream. He therefore steers toward a point above that which he wishes to reach, and so goes directly across.

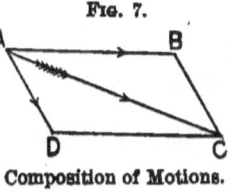

Composition of Motions.

—A bird, beating the air with both its wings, flies in a direction different from that which would be given by either one.

8. Composition of Forces.—When a body is thus acted on by two forces, we draw lines representing their directions, and mark off *AD* and *AB*, whose lengths represent their comparative magnitudes. We next complete the parallelogram and draw the diagonal *AC*, which denotes the *resultant* of these forces, and gives the direction in which the body will move.

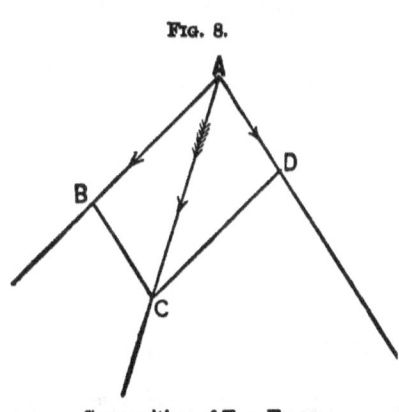

Composition of Two Forces.

28 MOTION AND FORCE.

If more than two forces act, we find the resultant of two, then of that resultant and a third force, and so on.

9. Triangle of Forces.—In Fig. 8 the resultant, AC, could have been obtained more easily by drawing AB to represent the magnitude and direction of one force, and then similarly BC for the other force. Connecting the initial point, A, of the first line with the terminal point, C, of the second line, we have AC for the magnitude and direction of the resultant, which completes a triangle.

10. Polygon of Forces.—Let F_1, F_2, F_3, F_4, and F_5 (Fig. 9), represent five forces acting on the same point at the same time. To find their resultant, we draw (Fig. 10) OA, AB, BC, CD, and DE, equal and parallel respectively to F_1, F_2, F_3, F_4, and F_5. Then, joining the first point with the last, we have OE to represent the magnitude and direction of their combined resultant. For OB is the resultant for OA and AB; OC for OB and BC; OD for OC and CD; and OE for OD and DE. This method is applicable to the representation of any number of forces.

Fig. 9.

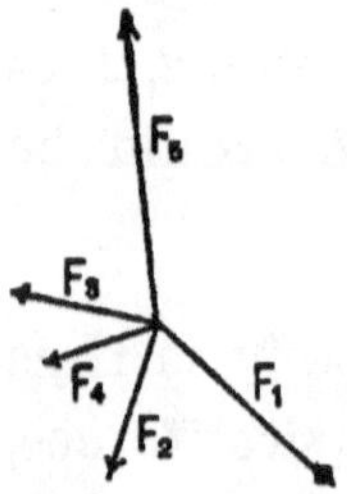

Five Forces acting at the same Point.

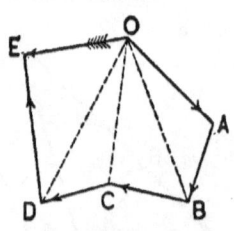

Fig. 10.

Polygon of Forces.

11. Resolution of Forces consists in finding what forces are equivalent to a given force under special conditions. A triangle is drawn, having the given

RESOLUTION OF FORCES.

force as one side.—*Example:* There is a wind, blowing nearly from the west (Fig. 11) against the sail, AB, of a vessel going northward. We may regard the windforce, WC, as the resultant of two forces, WD and DC. The former, being parallel to the sail, is not effective; the latter is perpendicular to it, and tends to drive the vessel nearly north-east. Again, resolving DC, we find this equivalent to two forces, DE and EC. The former pushes the vessel sideways, but is largely counteracted by the resistance of the water against the broad side; EC is in the direction of the ship's course, and propels it north.

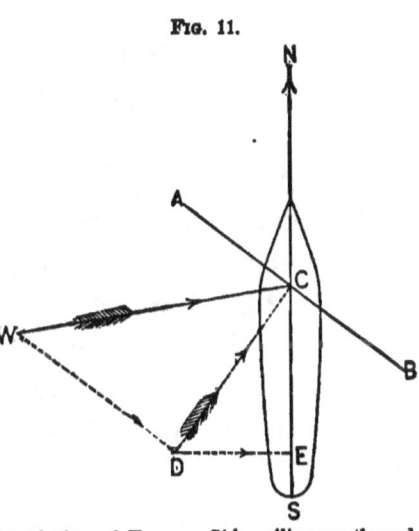

Fig. 11.

Resolution of Forces. Ship sailing northward.

By shifting the rigging, one vessel may sail into the harbor while another is sailing out, both driven by the same wind. In Fig. 12,

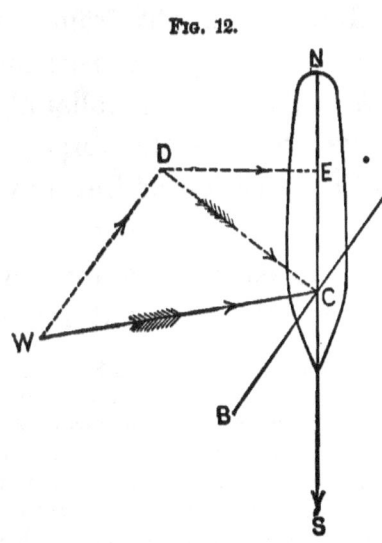

Fig. 12.

Resolution of Forces. Ship sailing southward.

which represents a ship sailing southward, the lettering and explanation is the same as for Fig. 11, if we substitute "south" for "north."* If the ship were required to go westward, it would *tack* alternately NW and SW. In this way its resultant direction might be almost in the "teeth of the wind."

A canal-boat drawn by horses is acted upon by a force which tends to bring it to the bank. This force may be resolved into two, one pulling toward the tow-path, and the other directly ahead. The former is counteracted by the shape of the boat and the action of the rudder; the latter draws the boat forward.

12. Motion in a Curve.—To change the direction of a moving body requires an external force (p. 22, § 4). When this force is a continuous force and the direction in which it acts differs continually from the path in which the body moves, the body will describe a curved line.—*Example:* A body thrown obliquely into the air describes a curve, because the force of gravity continually changes the direction of the moving body.

13. Circular Motion is produced when a moving body is drawn toward a center by a constant force.

---

* In a similar manner we may resolve the three forces which act upon a kite—viz., the pull of the string, the force of the wind, and its own weight. In Fig. 11, let AB represent the face of the kite. We can resolve WC, the force of the wind, into WD and DC. We next resolve DC into DE and EC. We then have a force, EC, which overcomes the weight of the kite and tends to lift it upward. The string pulls in the direction CD, perpendicularly to the face. The kite obeys neither one of these forces alone, but both, and so ascends in a direction CA between the two. It is really drawn up an inclined plane by the joint force of the wind and the string.

CIRCULAR MOTION.

Thus, when a sling is whirled, the stone is pulled toward the hand by the string, and as, according to the third law of motion, every action has its equal and opposite reaction, the hand is pulled toward the stone. If the string break, the stone will continue to move, according to the first law of motion, in a straight line in the direction of a tangent to the circle at that point. The tension of the string, acting inward, is called the *Centripetal* (*centrum*, the center, *petere*, to seek) force; and the reaction of the stone upon the string, acting outward, is termed the *Centrifugal* (*centrum*, the center, *fugere*, to flee) force.*

The following examples are among those usually

---

* It should be noticed that in circular motion there is but one true force concerned. It acts, however, upon a body in motion. The so-called centrifugal force has nothing to do with the production of the motion, being merely the resistance which the body offers by its inertia to the operation of the centripetal force, and ceases the instant that force is discontinued. It does not act at right angles to the centripetal force, as is often stated, but in direct opposition. A body never flies off from the center impelled by the centrifugal force, since that can never exceed the centripetal (action = reaction), and moreover the path of such a body is in the direction of a tangent, and not the radius of a circle. Thus, when water is thrown off a grindstone in rapid rotation, the tendency of the water to continue to move on in the direction of the straight line in which it is going at each instant (in other words, the inertia of the water) overcomes its adhesion to the stone, and it flies off in obedience to the first law of motion. So, also, when a grindstone, driven at a high speed, breaks, and the fragments are thrown with great velocity, we are not to suppose that the centrifugal force impels them through the air. That force existed only while the stone was entire. It was opposed to the force of cohesion, and in the moment of its triumph ceased, and the fragments of the stone fly off in virtue of the velocity they possess at that instant. Again, the so-called centrifugal force is not a real force urging bodies upward at the equator. The earth's surface is merely falling away from a tangent, and a part of the force of gravity is spent in overcoming the inertia of bodies. The term centrifugal force has caused much confusion, and will perhaps be discarded.

given to illustrate the action of the center-fleeing force: Water flies from a grindstone on account of the centrifugal force produced in the rapid revolution, which overcomes the *adhesion.*—In factories, grindstones are sometimes revolved with such velocity that this force overcomes that of *cohesion*, and the ponderous stones fly into fragments.—A pail full of water may be whirled around so rapidly that none will spill out, because the centrifugal force overcomes that of *gravity.*—When a horse is running around a small circle, he bends inward to overcome the centrifugal force.

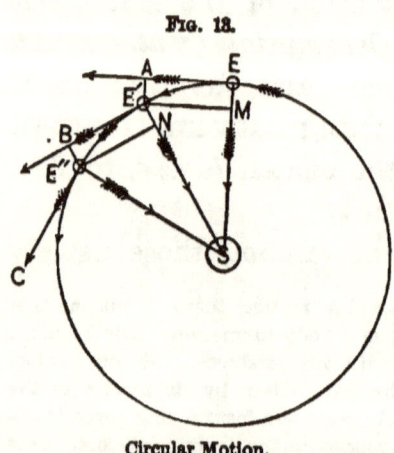

Circular Motion.

The heavenly bodies present the grandest example of circular motion. We may suppose the earth to have been moving originally in the direction *AE*. The attraction of the sun, however, drawing it in the direction *ES*, it passes along the line *EE'*. If the centripetal force were suddenly to cease, the earth would fly off into space along a tangent, as *EA*. The rapid revolution of the earth on its axis tends to throw off all bodies headlong. As this acts in opposition to gravity, it diminishes the weight of bodies at the equator, where it is greatest, being there equivalent to $\frac{1}{289}$ of the force of gravity. It also tends to drive the water on the earth from the poles toward the equator. Were the velocity of the earth's rotation to diminish, the water would flow back toward the poles, and tend to restore the earth to a spherical form.* This influence is well illustrated by the apparatus shown in Fig. 14. The hoop is made to slide upon its axis, and if revolved rapidly it will assume an oval form, bulging out more and more as the velocity is increased.†

Fig. 14.

14. Reflected Motion is produced by the reaction of a surface against which an elastic body is cast. If a perfectly elastic ball be thrown in the direction

---

* Since the earth's polar diameter is nearly twenty-seven miles shorter than its equatorial diameter, we are not sure that this motion of its waters would make it perfectly spherical.

† This apparatus is accompanied by objects to illustrate the principle that all bodies tend to revolve about their shortest diameters. "Tie to the middle of a lead-pencil a piece of string about three feet long. Suspend so that the pencil will balance itself. Now twist the end of the string between the thumb and the first finger of the right hand, steadying and holding the string with the left hand. A circular motion will thus be communicated to the pencil, and it will revolve around the point on which it is suspended. Tie a piece of white string around the middle of the pencil, or its center of gravity, simply to show the position of that point. Now tie the first piece of string half-way between the end of the pencil and the center of gravity, and communicate the circular motion described above, and we shall observe that the pencil will still revolve around the center of gravity, the point marked by the white string being at rest. It can thus be shown that any thing, of whatever shape, will tend to revolve on its shortest diameter. If the end links of a small steel chain (such as is often attached to purses or parasols) be hooked together, the string tied to a link, and the circular motion given, it will be observed that the chain begins to take an elliptical form, which gradually approaches that of a circle, until at last it becomes a circle, when it revolves horizontally. This shows that even a ring revolves on its shortest axis."

OP against the surface *AB*, it will rebound in the line *PQ*. The angle, *i*, between the direction *OP* and the perpendicular, *PR*, drawn at the point of incidence, is called the angle of incidence. The angle of reflection, *r*, is that between this perpendicular, *PR*, and the direction *PQ*. If *OP* represent the magnitude and direction of the

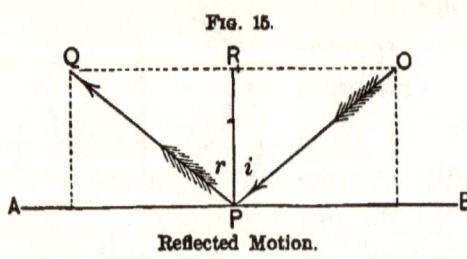

Fig. 15.

Reflected Motion.

incident force, it may be resolved into *OR* and *RP*. But the reaction, *PR*, is equal to the vertical portion, *RP*, of the incident force, while the horizontal portion is not checked. Hence *PQ* = *OP*, and the angle of incidence is equal to the angle of reflection.

15. Energy has been defined as the power of producing change of any kind (p. 4, § 5). To produce a change resistance must be overcome, and "work is done when resistance is overcome." Therefore energy may also be defined as the *power of doing work*. It is in general a power put into a body by means of work, and which comes out of it when it does work.—*Examples:* A wound-up clock, a red-hot iron. The difference between energy and momentum is easily illustrated. When a bullet is fired from a rifle, the momenta of both are equal, but the energy of the former, *i. e.*, its power of doing work, as piercing a board, is far greater. Energy is proportional to the square of the velocity of the moving body. Thus, a cannon-ball given double speed will penetrate four times as far into a wall; and a stone thrown upward at the rate of ninety-six feet per second will rise nine times as far as with a velocity of thirty-two feet.

16. Two Forms of Energy.—Energy may be either active or latent. When a rock is tumbling down a mountain-side, it exhibits the force of gravity in full sway; but when the rock was lodged on the mountain-top, it possessed the same energy, which could be developed at any moment by loosening it from its place. These two forms are known as energy of motion and energy of position, or kinetic and potential energy.*

17. Conservation of Energy.—The sum of all the energy in the universe remains the same while its transformations are infinite. One kind of energy is changed into another; from an available form to one that is not controllable. A hammer falls by the force of gravity. In coming to rest when stopped, it does the work of crushing what it hits, and its motion as a mass is converted into one of molecules, revealing

---

* The following may be taken as examples to show the difference between kinetic and potential energy. We wind a watch, and by a few moments of labor condense in the spring a potential energy, which is doled out for twenty-four hours in the kinetic energy of the moving wheels and hands. Lift a pendulum, and you thereby give the weight potential energy. Let it fall, and the potential changes gradually to kinetic. At the center of the arc the potential is gone and kinetic is possessed. Then the kinetic changes again to potential, which increases till the end of the arc is reached and the pendulum ceases to rise, when the energy is that of position, not of motion. Potential energy is like what is concealed, lying in wait and ready to burst forth on the instant. It is that of a loaded gun prepared for the arm of the marksman. It is that of a river trembling on the brink of a precipice, about to take the fearful leap. It is that of a weight wound up and held against the tug of gravity. It is that of the engine on the track with the steam hissing from every crevice. On the contrary, kinetic energy is that in actual operation. The bullet is speeding to the mark; the river is tumbling; the weight is falling; the engine is flying over the rails. It is that of heat radiating from our fires, and electricity carrying our messages over the continent.

itself to our touch as heat. The sun is continually sending forth radiant energy, which has been stored up in it by the aggregation of matter during untold ages. Its kinetic energy is thus becoming dissipated into potential energy; but, even after it ceases to glow, the grand total, including all that was once kinetic, will remain unchanged.

PRACTICAL QUESTIONS.

1. A rifle-ball thrown against a board standing edgewise, will knock it down; the same bullet fired at the board will pass through it without disturbing its position. Why is this?

2. Why can a boy skate safely over a piece of thin ice, when, if he should pause, it would break under him directly?

3. Why can a cannon-ball be fired through a door standing ajar, without moving it on its hinges?

4. Why can we drive on the head of a hammer by simply striking the end of the handle?

5. Suppose you were on a train of cars moving at the rate of 30 miles per hour; with what velocity would you be thrown forward if the train were stopped instantly?

6. In what line does a stone fall from the mast-head of a vessel in motion?

7. If a ball be dropped from a high tower, it will strike the ground a little east of a vertical line. Why is this?

8. It is stated that a suit was once brought by the driver of a light wagon against the owner of a coach for damages caused by a collision. The complaint was "the latter was driving so fast that when the two carriages struck, the driver of the former was thrown forward over the dashboard." On trial he was nonsuited, because his own evidence showed him to be the one who was driving at the unusual speed. Explain.

9. Suppose a train moving at the rate of 30 miles per hour: on the rear platform is a spring gun aimed parallel to the track and in a direction precisely opposite to the motion of the car. Let a ball be discharged with the exact speed of the train; where would it fall?

10. Suppose a steamer in rapid motion, and on its deck a man jumping. Can he jump farther by leaping the way the boat is moving than in the opposite direction?

11. Could a party play ball on the deck of an ocean steam-ship when

steaming along at the rate of 20 miles per hour, without making allowance for the motion of the ship?

12. Since action is equal to reaction, why is it not so dangerous to receive the "kick" of a gun as the force of the bullet?

13. If you were to jump from a carriage in rapid motion, would you leap directly toward the spot on which you wished to alight?

14. If you wished to shoot a bird in swift flight, would you aim directly at it?

15. At what parts of the earth is the centrifugal force least?

16. What causes the mud to fly from the wheels of a carriage in rapid motion?

17. What proof have we that the earth was once a soft mass?

18. On a curve in a railroad, one track is always higher than the other. Why is this?

19. What is the principle of the sling?

20. The mouth of the Mississippi River is about 2½ miles farther from the center of the earth than its source. In this sense it may be said to "run up hill." What causes this apparent opposition to the attraction of gravity?

21. Is it action or reaction that breaks an egg, when I strike it against the table?

22. Was the man philosophical who said that it "was not the falling so far, but the stopping so quick, that hurt him"?

23. If one person runs against another, which receives the greater blow?

24. Would it vary the effect if the two persons were running in opposite directions? In the same direction?

25. Why can you not fire a rifle-ball around a hill?

26. Why is it that a heavy rifle "kicks" less than a light shot-gun?

27. A man on the deck of a large vessel draws a small boat toward him. Can you express the ratio of the ship's motion to that of the boat?

28. Suppose a string, fastened at one end, will just support a weight of 25 lbs. at the other. Unfasten it, and let two persons pull upon it in opposite directions. How much can each pull without breaking it?

29. Can a man standing on a platform-scale make himself lighter by lifting up on himself?

30. Why can not a man lift himself by pulling up on his boot-straps?

31. With what momentum would a steam-boat weighing 1,000 tons, and moving with a velocity of 10 ft. per second, strike against a sunken rock?

32. With what momentum would a train of cars weighing 100 tons, and running 10 miles per hour, strike against an obstacle?

33. What would be the comparative kinetic energy of two hammers, one driven with a velocity of 20 ft. per second and the other 10 ft.?

34. If a 100 horse-power engine can propel a steamer 5 miles per hour, will one of 200 horse-power double its speed?

35. Why are ships becalmed at sea often floated by strong currents into dangerous localities without the knowledge of the crew?

MOTION AND FORCE.

36. A man in a wagon holds a 50-lb. weight in his hand. Suddenly the wagon falls over a precipice. Will he, while dropping, bear the strain of the weight?

37. Why are we not sensible of the rapid motion of the earth?

38. A feather is dropped from a balloon which is immersed in and swept along by a swift current of air. Will the feather be blown away or will it appear to drop directly down?

39. Suppose a bomb-shell, flying through the air at the rate of 500 ft. per second, explodes into two parts of equal weight, driving one half forward in the same direction as before, but with double its former velocity. What would become of the other half?

40. Which would have the greater penetrating power, a 10-lb. cannon-ball with a velocity of 1,000 ft. per second, or a 100-lb. ball with a velocity of 100 ft. per second?

41. There is a story told of a man who erected a huge pair of bellows in the stern of his pleasure-boat, that he might always have a fair wind. On trial, the plan failed. In which direction should he have turned the bellows?

42. If a man and a boy were riding in a wagon, and, on coming to the foot of a hill, the man should take up the boy in his arms, would that help the horse?

43. If we whirl a pail of water swiftly around with our hands, why will the water tend to leave the center of the pail?

44. Why will the foam collect at the hollow in the center?

45. If two cannon-balls, one weighing 8 lbs. and the other 2 lbs., be fired with the same velocity, which will go the farther?

46. Resolve the force of the wind which turns a common windmill, and show how one part acts to push the wheel against its support, and one to turn it around.

47. When an animal is jumping or falling, can any exertion made in mid-air change the motion of its center of gravity?

48. If one is riding rapidly, in which direction will he be thrown when the horse is suddenly stopped?

49. When standing in a boat, why, as its starts, are we thrown backward?

50. When carrying a cup of tea, if we move or stop quickly, why is the liquid liable to spill?

51. Why, when closely pursued, can we escape by dodging?

52. Why is a carriage or sleigh, when sharply turning a corner, liable to tip over?

53. Why, if you place a card on your finger and on top of it a cent, can you snap the card from under the cent, which will then drop on your finger?

54. Why is a "running jump" longer than a "standing jump"?

55. Why, after the sails of a vessel are furled, does it still continue to move? and why, after the sails are spread, does it require some time to get it under full headway?

56. Why can a tallow candle be fired through a board?

SUMMARY.

MATTER, so far as we know it, is in constant change. Change of place is termed motion. Terrestrial motion is restricted by friction, by the air, and by water. Friction is caused by the roughness of the surface over which a body moves. It may be decreased by the use of grease to fill up the minute projections, or by changing the sliding into rolling friction. Air and water must be displaced by a moving body; the resistance they offer is measured by the kinetic energy expended in overcoming it, and is hence proportional to the square of its velocity. Motion takes place in accordance with three laws; viz.: A moving body left to itself tends to go forever in a straight line with a constant velocity; a force has the same effect whether it acts alone or with other forces, and upon a body at rest or in motion; and action is equal and opposed to reaction. By means of the principles of the composition and resolution of forces, we can find the individual effect of a single force or the combined effect of several forces. To change the path of a moving body requires an external force. When the direction of the action of this force continually differs from the direction of the motion of the body, the path of the latter is that of a curved line. A particular case of this is circular motion. The path of a bullet or rocket in the air exemplifies curvilinear motion; and the movement of a stone whirled in a sling and of a planet revolving about the sun are illustrations of circular motion. When a rubber ball bounds back from a surface against which it is thrown, the angle of reflection equals the angle of incidence.

Energy is the power of overcoming resistance, *i.e.*, of doing work. This power, possessed by bodies, may be either latent—potential—as in the case of a suspended weight, or actual—kinetic—as in the case of the weight when falling.*

* When energy refers to that possessed by visible masses it is called *molar* energy; when it refers to the energy possessed by molecules it is called *molecular* energy. A body is hot because of the kinetic energy of its molecules. Heat is, therefore, an example of molecular energy. Light, sound, electricity are other forms of molecular energy. The different forms of energy are mutually convertible. The molar energy of a falling mass when stopped is transferred to its molecules and becomes the molecular energy of heat. This

HISTORICAL SKETCH.

ARISTOTLE taught that all motion is naturally circular, and this view was held by his school. He divided the phenomena of motion into two classes—the natural and the violent. As an instance of the former, he gave the falling of a stone, which constantly increases in velocity; and of the latter, a stone thrown vertically up, which being against nature, continually goes slower. Newton, in his "Principia," published in 1687, propounded the laws of motion as now received. Other philosophers, notably Galileo, Hooke, and Huyghens, had anticipated much of his reasoning, yet so slowly were his opinions accepted that "at his death," says Voltaire, "he had not more than twenty followers outside of England."

The law of the Conservation of Energy, Faraday, the great English physicist, pronounced "the grandest ever presented for the contemplation of the human mind." It has been established within the present century; yet we now know that former scholars had inklings of the wonderful truth. It arose in connection with discoveries on the subject of Heat, and its history will be treated of hereafter.

Consult Stewart's "Conservation of Energy"; Youmans' "Correlation of the Physical Forces"; Faraday's "Lectures on the Physical Forces"; Everett's "Deschanel's Natural Philosophy"; Tait's "Recent Advances in Physical Science"; Maxwell's "Matter and Motion"; "Appleton's Cyclopedia," Art. Correlation of Forces; Tyndall's "Crystalline and Molecular Forces," in Manchester Science Lectures, '73–4; Crane's "Ball Paradox," in Popular Science Monthly, Vol. X., p. 725.

energy in the steam-engine is reconverted into the molar energy of the train. Every change that takes place in the universe is a case either of transference or transformation of energy. Energy, though convertible, is as indestructible as matter—it is *conserved*. The principle of the Conservation of Energy teaches that the quantity of energy in the universe is constant, that it can neither be increased nor diminished. We cannot account for its origin; we only know its law of action, which we must finally refer to a Supreme Being.

III.

ATTRACTION.

"THE smallest dust which floats upon the wind
Bears this strong impress of the Eternal mind:
In mystery round it subtle forces roll,
And gravitation binds and guides the whole."

"Attraction, as gravitation, is the muscle and tendon of the universe, by which its mass is held together and its huge limbs are wielded. As cohesion and adhesion, it determines the multitude of physical features of its different parts. As chemical or interatomic action, it is the final source to which we trace all material changes."—ARNOTT.

ANALYSIS OF MOLECULAR FORCES.

ATTRACTION.
- I. MOLECULAR FORCES. —ATTRACTIVE AND REPELLENT FORCES.
  - 1. COHESION.
    1. Definition of Cohesion.
    2. Three States of Matter.
    3. Cohesion acts at Insensible Distances.
    4. Surface Tension of Liquids.
    5. Liquids Tend to Form Spheres.
    6. Solids Tend to Form Crystals.
    7. Annealing and Tempering.
    8. Rupert's Drop.
  - 2. ADHESION.
    1. Definition and Illustration of Adhesion.
    2. Capillarity.
       - (1.) *Direct.*
       - (2.) *Reversed.*
       - (3.) *Law of Capillarity.*
    3. Imbibition.
    4. Solution.
    5. Diffusion of Liquids.
    6. Diffusion of Gases.
    7. Osmose.
- II. ATTRACTION OF GRAVITATION.
  1. Law of Gravitation.
  2. Illustrations of Gravity.
  3. Mass and Weight.
  4. The Earth's Center of Gravity.
  5. The Center of Gravity of a Body.
  6. Laws of Weight.
  7. Falling Bodies.
     - (1.) *In a Vacuum.*
     - (2.) *In the Air, with Atwood's Machine.*
     - (3.) *Experiments with this Machine.*
  8. Equations of Bodies Falling Freely.
  9. Measurement of Kinetic Energy.
  10. Resistance of the Air to Falling Bodies.
  11. Equilibrium.
  12. The Pendulum.

ATTRACTION.

I. MOLECULAR FORCES.

Attractive and Repellant Forces.—If we take a piece of iron and attempt to pull it to pieces, we find that there is a force which holds the molecules together and resists our efforts. If we try to compress the metal, we find that there is a force which holds the molecules apart and resists our efforts as before. We thus see that there are two opposing forces—an attraction and a repulsion—which operate between the molecules and resist change of distance between them.

1. COHESION.

1. Cohesion is that force which holds together molecules of the same kind.

2. Three States of Matter.—Matter occurs in three states—*solid*, *liquid*, and *gaseous*. The molecules of a solid are held together by cohesion, and this force resists increase of distance between them. This resistance may, however, be overcome by the application of heat-energy. The molecules of a solid are not really fixed and at rest, but move swiftly to and fro within narrow limits, without being able to leave these limits. By the application of heat the path over which the molecules move and the intensity of their motion is increased, and the solid *expands*. At a certain stage in this process of heating the distance between the molecules becomes great enough to permit them to wander slowly through the substance. The solid has now become a *liquid*. Matter in this state has no independent form, but assumes the shape of the containing vessel. Continuing with the application of heat, the spaces between the molecules become finally so great that the attraction between them is canceled, permitting them to dart hither and thither with great velocity. This represents the *gaseous state*. In this state the molecules require to be restrained by the walls of the containing vessel. By pressure and cooling gases are liquefied; by further cooling liquids are rendered solid.

3. Cohesion Acts at Insensible Distances.—Take two bullets, and having flattened and cleaned one side of each, press them together with a twisting motion. They will cohere when the molecules are crowded into apparent contact.*—If two globules of mercury be brought near each other, at the instant they seem to touch they will suddenly coalesce.—Two freshly-cut surfaces of rubber, when warmed and pressed together, will cohere as if they formed one piece.—The process of welding illustrates this principle. A wrought-iron tool being broken, we wish to mend it. So we bring the iron to a white heat at the

---

* Surfaces may appear to the eye to be in contact when they are not actually so. Newton found, during some experiments on light (p. 220), that a convex lens or a watch-glass laid on a flat glass does not touch it, and can not be made to do so, even by a force of many pounds.

COHESION. 45

ends which we intend to unite. This partly overcomes the attraction of cohesion, and the molecules will move easily upon one another. Laying now one of the two heated ends upon the other, we pound them until the molecules are brought near enough for cohesion to grasp them.

4. Surface Tension of Liquids.—Within a liquid each molecule is pressed by the weight of all those above it, so that the slight cohesion between it and its neighbors is masked. At the surface there is no such liquid pressure. Cohesion causes the surface film to be in a state of tension like a sheet of stretched India rubber. A double film may be detached by using soapy water and lifting out of it the bowl of a pipe. The elasticity and toughness of the film are shown by blowing it into a bubble, whose surface is many times greater than that across the bowl that held it. A candle flame may be blown out by the bubble as it contracts toward its own center and expels a breeze of air through the tube of the pipe.*

5. Liquids Tend to Form Spheres.—Mix alcohol and water in such proportion that a drop of olive-oil

---

* There are many charming experiments that can be made with soap films. (See *Popular Science Monthly*, Vol. IX., p. 575; *Scientific American*, May 15, 1886; *Scientific American Supplement*, Jan. 25, 1879.) A good recipe for making soap solution is as follows: Procure some of the best white Castile, or palm-oil soap. Scrape from it about four ounces of thin shavings, put these into a quart bottle of purest rain-water, or distilled water, and shake until the strongest clear solution of the soap is had. Then add a pint of pure concentrated glycerine. A film from this mixture will last for hours, and bubbles over a foot in diameter are easily made.

will sink into it without going to the bottom. The action of gravity on it is just balanced by the buoyancy of the liquid, so that cohesion can act without much interference. Like a soap-bubble, the outside film of the oil tends to contract upon its interior, and a nearly perfect sphere remains suspended. The contraction of the tough surface film is what produces the roundness of dew-drops, rain-drops, globules of mercury, and melted lead as it falls and cools into round shot.

6. Solids Tend to Form Crystals.—A liquid in becoming solid shows a tendency to assume a regular form bounded by plane surfaces termed a *crystal*.

Fig. 16.

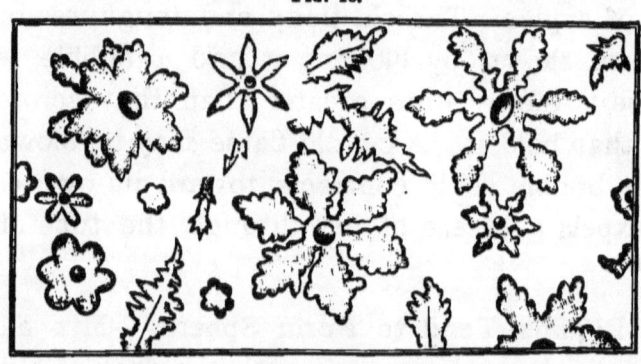

Ice Flowers.

Crystals of different substances show in most cases marked differences in form, which may serve for the recognition of the substance.* When different sub-

---

* Epsom salt crystallizes in four-sided prisms, common salt in cubes, and alum in octahedra. We can illustrate the formation of the last by adding alum to hot water until no more will dissolve. Then suspend strings across the dish and set it away to cool. Beautiful octahedral crystals will

stances are contained in the same solution, they separate on crystallization, and each molecule goes to its own. The exquisite beauty of these crystalline forms is seen in snow-flakes and the frost-work traced on a cold morning upon the windows or the stone-flagging. A beam of light passed through a block of ice reveals these crystals as a mass of star-like flowers (Fig. 16).*

Melted iron rapidly cooled in a mold has not time to arrange its crystals. If, however, the iron be afterward violently jarred, as when used for cannon, rail-cars, etc., the molecules take on the crystalline form and the metal becomes brittle.†

7. Annealing and Tempering. — If a piece of wrought-iron be heated and then plunged into water, it becomes hard and brittle. If, on the contrary, it be heated and cooled slowly, it is made tough and flexible. Steel is tempered by heating red-hot, then cooling quickly, and afterward re-heating and cooling slowly. It becomes then one of the most elastic and tough of known substances.

8. The Rupert's Drop is a tear of melted glass dropped into water, and cooled quickly. The exte-

---

collect on the threads and sides of the vessel. The slower the process, the larger the crystals.

* It is noticeable that, as the crystals melt, at the center of each liquid flower is a vacuum, showing that there is not enough water formed to fill the space occupied by the crystal, and that the solid contracts as it passes into a fluid (p. 271). This experiment is easily tried. The ice must be cut parallel to the plane of its freezing and be not over half an inch thick. A common oil lamp will furnish the light.

† On examining such a piece of iron, which can easily be procured at a car or machine shop, we can see in a fresh fracture the smooth, shining faces of the crystals.

rior at once becomes rigid, while the interior is still hot and expanded. When the whole mass is cool, the interior is in a state of strain. If the tail of the drop be nipped off, so that the exterior shell is broken, the tension will cause the mass to fly into powder with a sharp explosion. All glassware, when first made, is brittle, but it is annealed by being drawn slowly through a long oven, highly heated at one end, but quite cool at the other. During this passage, the molecules of glass have time to arrange themselves in a stable position.*

FIG. 17.

Rupert's Drop.

PRACTICAL QUESTIONS.

1. Why can we not weld a piece of copper to one of iron?
2. Why is a bar of iron stronger than one of wood?
3. Why may a piece of iron, when perfectly welded, be stronger than before it was broken?
4. Why do drops of different liquids vary in size?
5. When you drop medicine, why will the last few drops contained in the bottle be of a larger size than the others?
6. Why are the drops larger if you drop them slowly?
7. Why, if you melt scraps of lead, will they form a solid mass when cooled?
8. In what liquids is the force of cohesion greatest?
9. Why does iron, by continued jarring, become brittle?
10. Why can glass be welded?
11. Name some substances that can not be welded. Why not?
12. What liquids would you select for showing surface tension?

---

* "The restoration of cohesion is beautifully seen in the gilding of china. A figure is drawn upon the china with a mixture of oxide of gold and an essential oil. The article is then heated, whereby the essential oil and the oxygen of the gold are expelled, and a red-brown pattern remains. This consists of pure gold in a finely-divided state, without luster. By rubbing with a hard burnisher, the particles of gold cohere and reflect the rich yellow color of the polished metal."

2. ADHESION.

1. Adhesion is the force which holds together molecules of different kinds.—*Examples:* Two pieces of wood are fastened together with glue, two pieces of china with cement, two bricks with mortar, two sheets of paper with mucilage, and two pieces of tin with solder.—Syrup and coal-oil are purified by filtering through animal charcoal.—Bubbles can be blown from soap-suds, because the soap by its adhesive force holds together the particles of water.

2. Capillarity.—If there is strong adhesion between a liquid and a solid partly immersed in it, the liquid rises above the general level and wets the solid, causing the surface film along the line of contact to be concave upward. This is shown in Fig. 18, which represents a tube of glass dipped in water. On the inner side of the tube the concavity is more marked in proportion as the bore is less. The contraction of the surface film (p. 45) balances the weight of the liquid above the level within the tube. This is called capillarity (*capillus*, a hair), because best shown in tubes with a bore as fine as a hair.

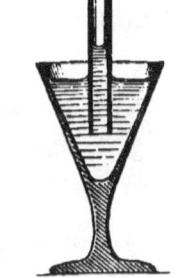

Fig. 18.

Direct Capillarity.

Fig. 19.

Reversed Capillarity.

If the tube be thrust into a liquid, like mercury, which does not wet it, the capillarity is reversed; the liquid is convex upward, and within the tube it is depressed.

A striking illustration of the effect of narrowing

50   ATTRACTION.

the exposed surface of the film may be seen by putting two clean glass plates edgewise into colored water so that their lower part shall be immersed, with a pair of upright edges touching (Fig. 20.) Just at the edge the liquid rises to the top; the height decreases as the successive distances between the two plates increases, and the liquid surface seen edgewise forms a curve called the Hyperbola.

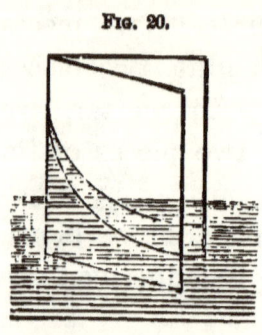

Fig. 20.

The Capillary Curve.

Law of Capillarity.*—The height to which a liquid rises in a wetted tube varies inversely as the diameter.—*Example:* In a tube whose inner diameter is 1 mm. (see p. 12), water rises at ordinary temperature (62°F) to a height of 29 mm. (a little over an inch); if the tube be only ½ mm. in diameter, the height will be twice as great. Cold water rises higher than warm water, or than alcohol or ether.

3. Imbibition.—Many porous bodies (*sensible* pores, p. 8), such as sugar, blotting paper, sand, a lamp wick, a towel, absorb liquids at a rate which is increased if the materials be warm.† This is due to the

---

* For further discussion of Surface Tension and Capillarity, consult Maxwell's "Theory of Heat," p. 280; "American Journal of Science," Dec., 1882, p. 416; Pickering's "Physical Manipulation," Vol. I., p. 102; Daniells' "Physics," p. 245; Deschanel's "Natural Philosophy," p. 130.

† In the same way, water is drawn to the surface of the ground to furnish vegetation with the materials of growth. Even in the winter, when the surface is frozen, the water still finds its way upward, and freezes into ice, which in the spring produces mud, although there may have been little rain or snow.

ADHESION. 51

attraction between the solid and the liquid. Ropes absorb water by imbibition, swell, and shrink often to breaking.*

4. Solution.—Sugar will dissolve in water, because the adhesion between the two substances is stronger than the cohesion of the sugar.† As heat weakens cohesion, it hastens solution, so that a substance generally dissolves more rapidly in hot water than in cold. In like manner, pulverizing a solid aids solution. Liquids also absorb gases by adhesion. Thus water contains air, which renders it pleasant to the taste. As pressure and cold weaken the repellent force, they favor the adhesion between the molecules

---

* It is 1586. The Egyptian obelisk, weighing a million pounds, is to be raised in the square of St. Peter's, Rome. Pope Sixtus V. proclaims that no one shall utter a word aloud until the engineer announces that all danger is passed. As the majestic column ascends, all eyes watch it with wonder and awe. Slowly it rises, inch by inch, foot by foot, until the task is almost completed, when the strain becomes too great. The huge ropes yield and slip. The workmen are dismayed, and fly wildly to escape the impending mass of stone. Suddenly a voice breaks the silence. "*Wet the ropes*," rings out clear-toned as a trumpet. The crowd look. There, on a high post, standing on tiptoe, his eyes glittering with the intensity of excitement, is one of the eight hundred workmen, a sailor named Bresca di S. Remo. His voice and appearance startle every one; but his words inspire. He is obeyed. The ropes swell and bite the stone. The column ascends again, and in due time stands securely on its pedestal. The daring sailor is not only forgiven, but his descendants to this day enjoy the reward of providing the palm-branches used on Palm Sunday at St. Peter's.

† This contest between adhesion and cohesion is seen when we let fall on water a drop of oil. Adhesion tends to draw the oil to the liquid, so as to mix thoroughly, and cohesion to prevent this. The extent to which the drop will spread will depend on the relation of the two attractions, and vary for every substance. Thus each oil has its own COHESION FIGURE, which enables the chemist readily to detect differences and mixtures. —*Experiments:* Dissolve a little salt in a glass of water, and touch the surface of the liquid with a pen full of ink. The characteristic figures will quickly appear.—Dissolve in water a pinch of salt and a lump of loaf-sugar. Touch the surface with lunar caustic. The figure of nitrate of silver will be seen.

of a gas and water. Soda-water receives its effervescence and pungent taste from carbonic-acid gas, which, being absorbed under great pressure, escapes in sparkling bubbles when the pressure is removed.

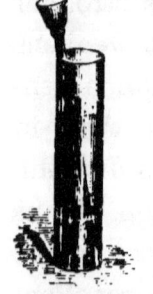
Fig. 21. Diffusion of Liquids.

5. Diffusion of Liquids.—Let a jar be partly filled with water colored by blue litmus. Then, by a funnel-tube, pour clear water containing sulphuric acid to the bottom, beneath the colored water. At first, the two will be distinctly defined, but in a few days they will mix, as will be seen by the change of color from blue to red. A drop of sulphuric acid may thus be distributed through a quart of water. Most liquids will mingle when brought in contact. If, however, there is no adhesion between their molecules, they will not mix, and will separate even after having been thoroughly shaken together. For example, shake together mercury, water, and sweet oil; they will soon separate and settle in layers with the oil at the top and the mercury at the bottom. Diffusion is a slow process, so we generally help it by shaking or stirring the mixture of liquids.*

Fig. 22. Diffusion of Gases.

6. Diffusion of Gases.—Hydrogen gas is only $\frac{1}{14}$ as heavy as common air. Yet, if two bottles be arranged as in Fig. 22, the lower one filled with the

---

* A story is told of some negroes in the West Indies who supplied themselves with liquor by inverting the neck of a bottle of water in the bung-

ADHESION. 53

heavy gas, and the upper with the lighter, the gases will soon be uniformly mixed.*

7. Osmose.—When two liquids are separated by a thin substance, the interchange may be modified in

hole of a cask of rum. The water sank into the barrel, while the rum rose to take its place. Water and rum diffuse readily, but rum is lighter and requires time to diffuse to the bottom.

* This phenomenon is explained by the theory that the molecules of all bodies are in rapid motion (p. 44, § 2). In a gas the molecules are assumed to move in straight lines, collide with each other, rebound, and so have their directions continually changed. The paths over which they move are inconceivably minute, but the velocity with which they move is correspondingly great. The particles of ammonia gas, for example, are flying to and fro at the rate of twenty miles per minute. "Could we, by any means," says Prof. Cooke, "turn in one direction the actual motion of the molecules of what we call still air, it would become at once a wind blowing seventeen miles per minute, and exert a destructive power compared with which the most violent tornado is feeble."—Invert a bottle over a lighted candle, and the oxygen of the inclosed air being soon consumed the flame goes out. Instead of the bottle, use a foolscap-paper cone. There will be an interchange of gases through the pores of the paper and the light will burn with moderate freedom.

Diffusion of Gases is still more strikingly shown in the experiment of Fig. 23, devised by Prof. Graham. Fit a porous cup used in Grove's Battery (p. 320) with a cork and glass tube. Fasten the tube so that it will dip beneath the colored water in the glass. Then invert over the cup a jar of hydrogen. The gas will pass through the sensible pores of the earthenware and down the tube so rapidly, as almost instantly to bubble up through the water. Rose balloons lose their buoyancy, because the hydrogen escapes through the pores of the rubber. If they were filled with air and placed in a jar of hydrogen, that gas would creep in so rapidly as to burst them.—In performing the experiment shown in Fig. 23, coal-gas may be used. After the bubbling ceases, on withdrawing the jar water rises in the tube. The thin gas diffuses out into the denser air, producing a partial vacuum within.

FIG. 23.

Graham's Diffusion Tube.

a curious manner, according to the nature of the liquid and the substance used. At the end of a glass tube (Fig. 24) fasten a bladder filled with alcohol.

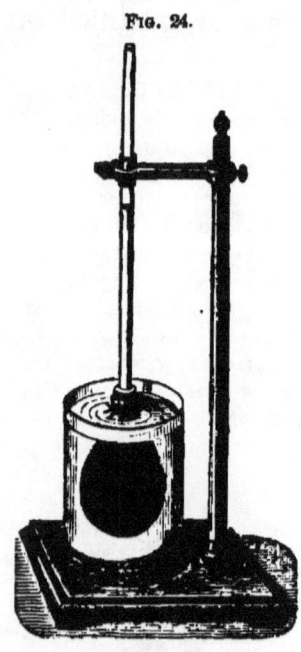

Fig. 24.

Osmose.

Insert into a jar of water, and mark the height to which the alcohol ascends in the tube. The column will soon begin to rise slowly. On examination, we shall see that the alcohol is passing out through the pores of the bladder and mixing with the water, while the water is coming in more rapidly. The bladder is not porous in the sense of having sensible pores.

Diffusion of fluids, through the medium of a substance which attracts them unequally, is called osmose. It is applied by the chemist in methods of analysis where the separation of substances is based on their unequal diffusibility. Crystals in solution pass readily through animal membrane, like bladder, while substances that do not crystallize, like gum, gelatine, or white of egg, are stopped.

PRACTICAL QUESTIONS.

1. Why does cloth shrink when wet?
2. Why do sailors at a boat-race wet the sails?
3. Why is writing-paper sized?
4. Why does paint prevent wood from shrinking?

5. What is the shape of the surface of a glass-full of water? Of mercury?
6. Why can we not perfectly dry a towel by wringing?
7. Why will not water run through a fine sieve when the wires are greased?
8. Why will camphor dissolve in alcohol, and not in water?
9. Why will mercury rise in zinc tubes as water will in glass tubes?
10. Why will ink spilled on the edge of a book extend farther inside than if spilled on the side of the leaves?
11. If you should happen to spill some ink on the edge of your book, ought you to press the leaves together?
12. Why can you not mix water and oil?
13. What is the object of the spout on a pitcher? *Ans.* The water would run down the side of the pitcher by the force of adhesion, but the spout throws it into the hands of gravitation before adhesion can catch it.
14. Why will water wet your hand, while mercury will not?
15. Why is a pail or tub liable to fall to pieces if not filled with water or kept in a damp place?
16. Name instances where the attraction of adhesion is stronger than that of cohesion.
17. Why does the water in Fig. 18 stand higher inside of the tube than next the glass on the outside?
18. Why will clothes-lines tighten and sometimes break during a shower?
19. In casting large cannon, the gun is cooled by a stream of cold water. Why?
20. Why does paint adhere to wood? Chalk to the blackboard?
21. Why does a towel dry one's face after washing?
22. Why will a greased needle float on water?
23. Why is the point of a pen slit?
24. Why is a thin layer of glue stronger than a thick one?

II. ATTRACTION OF GRAVITATION.

WE have spoken of the attraction existing between the molecules of bodies at minute distances. We now notice an attraction which acts at all distances.

1. Law of Gravitation.—Hold a stone in the hand, and you feel a power constantly drawing it to the ground. We call this familiar phenomenon *weight*. It is really the attraction of the earth pulling the stone back to itself—an instance of a general law, one operation of an ever-active force. For *every particle of matter in the universe* attracts every other particle with a force proportional to the product of their masses, and increasing as the square of the distance decreases.*

Gravitation is the general term applied to the attraction that exists between all bodies in the universe. *Gravity* is the earth's attraction for terrestrial bodies; it tends to draw them toward the center of the earth.

Fig. 25.

Deflection of a Plumb-line by a mountain. (Exaggerated.)

2. Illustrations of Gravity.—A stone falls to the ground because the earth attracts it; but in turn the stone attracts the earth. Each moves to meet the other, but the stone passes through as much greater distance than the earth as its mass is less. The mass of the earth is so great that its motion is imperceptible.—A plumb-line hanging near a mountain is attracted from the vertical. In Fig. 25, AB represents the ordinary posi-

---

* The force of gravitation acts on every particle of matter, and hence it is not confined to our own world. By its action the heavenly bodies are bound to one another, and thus kept in their orbits. It may help us to conceive how the earth is supported, if we imagine the sun letting down a huge cable, and every star in the heavens a tiny thread, to hold our globe in its place, while it in turn sends back a cable to the sun and a thread to every one of the stars. So we are bound to them and they to us. Thus the worlds throughout space are linked together by these cords of mutual attraction, which, interweaving in every direction, make the universe a unit.

tion of the line, while AC indicates the attractive power (greatly exaggerated) of the mountain.*

3. Mass and Weight.—The quantity of matter in a body is its mass. The measure of the earth's attraction upon it is its weight. If m be the number of units of mass in a body, and g be the number of units of force expressing the earth's attraction, then its weight, w, is equal to m multiplied by g; or, $w = mg$, whence

$$m = \frac{w}{g}.$$

4. The Earth's Center of Gravity is that point within it where the attraction of all the particles on any one side is equal to the attraction of all those on the opposite side. As an attracting mass the whole earth may be regarded as if it were concentered at this point. Its position is probably very near the geometric center. When the earth's center is mentioned we generally mean its center of gravity.

5. The Center of Gravity of a Body is that point about which it may be balanced. A straight line from this point to the earth's center of gravity is called the line of direction. It is also called a vertical, or plumb-line. Gravity tends to cause the body to move along this line toward the center.†

---

* Maskelyne, in 1774, found the attraction of Mount Schehallien to deflect a plumb line 12″. By comparing this force with that of the earth, the specific gravity of the earth was estimated to be five times that of water. Later investigations make it 5.67.

† *Downward* means toward the earth's center; *upward* means the opposite. Any two bodies moving downward, one from America and the other from Europe, if unresisted, would meet at the earth's center if their fall were properly timed.

ATTRACTION.

6. Laws of Weight.—I. *The weight of a body at the center of the earth is nothing;* for since the opposite attractions are mutually balanced there can be no tendency to motion in any direction.

II. *The weight of a body above the surface of the earth decreases as the square of the distance from the center of the earth increases.**

III. *The weight of a body varies on different portions of the surface of the earth.*† It will be least at the equator, because (1), on account of the bulging form of our globe a body is there farther from the earth's center; and (2), the centrifugal force is there strongest. It will be greatest at the poles, because (1), on account of the flattening of the earth a body at a pole is there nearer to the earth's center; (2), there is no centrifugal force at the poles.

7. Falling Bodies.—I. *Under the influence of the constant force of gravity alone all bodies fall with equal rapidity.*

---

* A body at the surface of the earth (4,000 miles from the center) weighs 100 lbs. What would be its weight 1,000 miles above the surface (5,000 miles from the center)? SOLUTION: (5,000 mi.)² : (4,000 mi.)² :: 100 lbs. : x = 64 lbs. Or, its weight would decrease in the ratio of $\frac{4000^2}{5000^2}$ = $\frac{16}{25}$. Hence it would weigh $\frac{16}{25}$ × 100 lbs. = 64 lbs.—The weight of a body below the surface of the earth is commonly said to decrease directly as the distance from the center decreases. Thus, 1,000 miles below the surface, a body would lose ¼ its weight. In fact, however, the density of the earth increases so much toward the center, that "for $\frac{1}{5}$ of the distance the force of gravity actually becomes stronger than on the surface."

† In these statements concerning weight, a spring-balance is supposed to be employed. If it be graduated to indicate correctly at a medium latitude, it would show too little at the equator, and too much at the poles. In other words, a pound weighed with such a spring-balance at the equator would contain a greater mass of matter than one weighed at the poles by about $\frac{1}{192}$ part.

FALLING BODIES. 59

This is well illustrated by the "guinea and feather experiment." Let a coin and a feather be placed in a tube, and the air exhausted. Quickly invert the tube, and the two bodies will fall in nearly the same time. Let in the air again, and the feather will flutter down long after the coin has reached the bottom.* Hence we conclude that in a vacuum all bodies descend with equal velocity, and that the resistance of the air and the adhesion of the feather to the tube are the causes of the variation we see between the falling of light and of heavy bodies in it.

II. ATWOOD'S MACHINE.†—To deduce the laws of falling bodies we make use of Atwood's machine (Fig. 27). This consists of a very light grooved wheel, w,

FIG. 26.

Guinea and Feather Experiment.

---

* The same fact may be noticed in the case of a sheet of paper. When spread out, it merely flutters to the ground; but when rolled in a compact mass, it falls quickly. In this case we have not increased the force of attraction, but we have diminished the resistance of the air. "It is difficult for many pupils to understand how, under the influence of gravity alone, all bodies fall with equal rapidity. An illustration, which is usually effective, is that of a number of bodies of the same kind, say bricks, which will separately fall in the same space of time. The pupil will admit that, if all of them are connected together, inasmuch as nothing is thereby added to their weight, there is no reason why the mass of bricks should not fall in the time of a single one, notwithstanding it is a larger body."—WM. H. TAYLOR.

† If the teacher is not provided with Atwood's machine, or if the pupils

ATTRACTION.

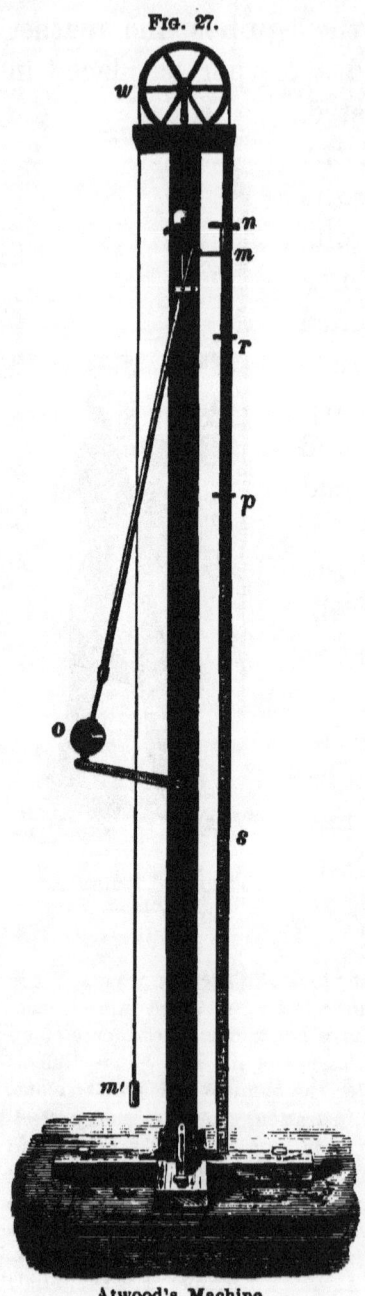

FIG. 27.

Atwood's Machine.

pivoted at the top of a firm vertical pillar, on one side of which is a graduated strip, s, divided into inches or centimeters. A silk thread, passing over the wheel, supports two equal masses, m and m'. The force of gravity on one of these just balances its force on the other. A small cross-bar, n, placed upon m, gives vertical motion, by the action of gravity on it, to the whole mass, which we may call M, made up of $(m + m' + n)$. Since the momentum of m, moving downward, is balanced by that of m', moving upward, the rate of motion of M is as much less than if it were falling freely as the mass of the cross-bar is less than

are unfamiliar with algebra, it may be found best to omit this discussion of Falling Bodies. The subject receives special attention in the present edition on account of the united request of many teachers. It has been reduced within the narrowest limit consistent with clearness.

FALLING BODIES. 61

the whole mass, M. An allowance has to be made for the mass of the wheel, which is put in circular motion at the same time. By properly choosing the mass of the cross-bar, the motion of M may be made so slow that the distance through which it passes in each succeeding second may easily be measured on the graduated strip along which m falls. A pendulum, o, marks the successive seconds, and is so arranged as to release the support of m at the proper moment, thus allowing M to move. Attached to the graduated strip are a movable ring, r, and a movable plate, p. If these be placed as shown in Fig. 27, the cross-bar would be caught by the ring, while m passing through would continue to move until stopped by the plate p.

EXPERIMENTS WITH ATWOOD'S MACHINE.—Suppose the mass of the cross-bar be so adjusted to that of m and m' that when m is released by the pendulum it will pass through 5 cm. (p. 4) during the first second. Let the plate, p, be removed, and the ring, r, be placed 5 cm. below the starting-point of m. Then allowing the system M to fall, the cross-bar will be caught by the ring exactly at the end of one second and stopped; the remainder of the system, however, will continue its motion (p. 22, § 4). m passing through the ring will be seen to pass over 10 cm. in each successive second until it reaches the foot of the instrument. 10 cm. per second, then, is the velocity produced by the force of gravity acting for *one* second on the cross-bar. Let the ring, for a second experiment, be placed 20 cm. below the starting-

62   ATTRACTION.

point. On allowing the system to fall, the cross-bar will now be caught exactly at the end of two seconds. *m* passing through the ring will be seen to move with a velocity of 20 cm. per second. We have evidently doubled the velocity by allowing the force of gravity to act for two seconds on the cross-bar. Similarly, by allowing gravity to act for three seconds, the velocity will be trebled. The longer, therefore, the force is allowed to act the greater will be the velocity produced. The distances passed over *during* the successive seconds were 5, 15, 25 cm. (see Fig. 28). *Under the action of a constant force, therefore, a body will move with a continually increasing velocity.* The gain in velocity per unit of time is called the *acceleration*, and the motion itself is described as *accelerated motion*. If the acceleration for successive seconds, as in our experiments, has the same numerical value, the motion is called *uniformly accelerated motion*, and this is the motion of a falling body.

From the first experiment it is seen that during the first second the system *M* passed over 5 cm., but that the acceleration

ATTRACTION. 63

produced in that time was double this, or 10 cm. per second. *A body, therefore, under the action of a constant force and starting from a state of rest passes over a distance during the first second equal to one half the acceleration produced during that time.*

On repeating the experiments with cross-bars of greater mass, it will be found that the accelerations produced increase with the mass of the cross-bar, that is, with the force employed. *A constant force may, therefore, be measured by the acceleration it imparts to unit mass* (page 23).

EQUATIONS OF FALLING BODIES.—These results are combined in the accompanying table, where the acceleration, 10 cm., is represented by f.

FIG. 29.

t	1	2	3	4		
v	f	$f \times 2$	$f \times 3$	$f \times 4$	$v = ft$	(1)
s	$\tfrac{1}{2}f$	$\tfrac{1}{2}f \times 3$	$\tfrac{1}{2}f \times 5$	$\tfrac{1}{2}f \times 7$	$s = \tfrac{1}{2}f(2t-1)$	(2)
S	$\tfrac{1}{2}f$	$\tfrac{1}{2}f \times 4$	$\tfrac{1}{2}f \times 9$	$\tfrac{1}{2}f \times 16$	$S = \tfrac{1}{2}ft^2$	(3)

The horizontal line marked t gives the time in seconds; that marked v, the velocity at the end of each second; that marked s, the space traversed during each successive second; and that marked S, the whole space traversed from the beginning of fall to the end of each second. Taking any one of the vertical columns, such as the fourth, we see that

(1.) At the end of any given second, the velocity

is equal to the acceleration multiplied by the number of the second; or, $v = ft$.

(2.) During any particular second, the space traversed is equal to half the acceleration multiplied by one less than double the number of the second; or, $s = \tfrac{1}{2}f(2t - 1)$.

(3.) The whole space traversed during a given number of seconds is equal to half the acceleration multiplied by the square of the time in seconds; or, $S = \tfrac{1}{2}ft^2$.

8. Bodies Falling Freely.—If a body falls freely, the acceleration is about 9.8 meters, or 32 feet.* If we represent this by g in the equations just deduced, we have

$$v = gt \quad \ldots \ldots \ldots \ldots \ldots (1)$$

$$s = \tfrac{1}{2}g(2t - 1) \quad \ldots \ldots \ldots (2)$$

$$S = \tfrac{1}{2}gt^2 \quad \ldots \ldots \ldots \ldots (3)\dagger$$

When a body is thrown upward, gravity causes it to lose 32 feet in velocity each second until it ceases to ascend. The velocity with which it begins to rise, and its time of rising, must be the same as the velocity with which it ends its fall, and the time of falling. The laws of falling bodies may hence be applied.

---

* For the latitude of New York, $g = 32.16$ more nearly, or 980.2 cm. This varies slightly with the latitude and elevation of a place; but for ordinary problems it will be sufficient to assume the values given in the text.

† An additional formula that is often useful may be obtained by eliminating t in combining equations (1) and (3). The result is $v = \sqrt{2gS}$. Since S represents height (h), it is often expressed $v^2 = 2gh$. (4).

EQUILIBRIUM.

9. Measurement of Kinetic Energy. (See p. 34.) —To lift a body through any height energy must be expended. The measure of this is the weight (w), multipled by the height (h), to which it is lifted. Using the initial letters of these words for symbols, we have

$$K = wh.$$

But from the equation, at the bottom of p. 64, we have $h = \dfrac{v^2}{2g}$. And from p. 57, $w = mg$. Substituting these values and reducing, we have

$$K = \tfrac{1}{2}mv^2.$$

10. Resistance of the Air to Moving Bodies.—A body in falling or otherwise moving through the air expends energy in overcoming the resistance of the air. From the formula just deduced we see that this is proportional to the mass of air moved and to the square of the velocity. The flight of a cannonball is never so great as it would be if shot through a vacuum. Practically it is not easy to calculate beforehand the amount of energy to be lost through resistances.

11. Equilibrium.—When a body is at rest the forces which act on every molecule in it are said to balance one another, or to be in equilibrium. The most important of these forces is gravity.

(1.) THREE STATES OF EQUILIBRIUM.—1st. A body is in *stable equilibrium* when the center of gravity is below the point of support, or when any movement tends to raise the center of gravity. In Fig. 30, the cork and two knives together form a connected body whose center of gravity is outside, just beneath the needle. By pushing either knife a few oscillations are produced, but a position of rest is soon recovered. Any movement of the toy shown in Fig. 31 tends to raise the center of gravity, and it returns quickly to a state of rest.

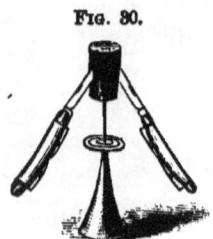
Fig. 30.
Stable Equilibrium.

2d. A body is said to be in *unstable equilibrium* when the center of gravity is above the point of support, or when any movement tends to lower the center of gravity. If we take the cork as arranged with the knives in Fig. 30, and invert it, we shall have difficulty in balancing the needle; and, if we succeed, it will readily topple off, as the least motion tends to lower the center of gravity.

Fig. 31.
Stable Equilibrium.

3d. A body is said to be in *indifferent equilibrium* when the center of gravity is at the point of support, or when any movement tends neither to elevate nor lower the center of gravity. A ball of uniform density on a level surface will rest in any position, because the center of gravity moves in a line parallel to the floor.

(2.) GENERAL PRINCIPLES.—(*a.*) The center of gravity tends to seek the lowest point.

(*b*) A body will not tip over while the line of direction falls within the base, but will as soon as it falls without.*

(*c.*) In general, narrowness of base combined with height of center of gravity, tends to instability;† breadth of base and lowness of center of gravity, produce stability.

(3.) PHYSIOLOGICAL FACTS.—Our feet and the space between them form the base on which we stand. By turning our toes outward, we increase its breadth.

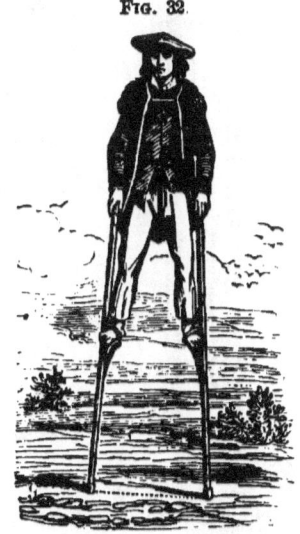

FIG. 32.

Walking on Stilts.

---

* The Leaning Tower of Pisa, in Italy, beautifully illustrates this principle. It is about 188 feet high, and its top leans 15 feet, yet the line of direction falls so far within the base that it is perfectly stable, having stood for seven centuries. The feeling experienced by a person who for the first time looks down from the lower side of the top of this apparently impending structure is startling indeed.

† "This is shown by the difficulty in learning to walk upon stilts. The art of balancing one's self may, however, be acquired by practice, as is seen in the Landes of south-western France. During a portion of the year these sandy plains are half covered with water, and in the remainder are still very bad walking. The natives accordingly double the length of their legs by stilts. Mounted on these wooden poles, which are put on and off as regularly as the other parts of their dress, they appear to strangers as a new and extraordinary race, marching with steps of six feet in length, and with the speed of a trotting-horse. While watching their flocks, they support themselves by a third staff behind, and then with their rough sheep-skin cloaks and caps, like thatched roofs, seem to be little watch-towers, or singular lofty tripods, scattered over the country."—ARNOTT.

When we stand on one foot, we bend over so as to bring the line of direction within this narrower base. When we walk, we incline to the right and the left alternately. When we walk up hill we lean forward, and in going down hill we incline backward, in unconscious obedience to the laws of gravity. We bend forward when we wish to rise from a chair, in order to bring the center of gravity over our feet. In walking we lean forward, so as bring the center of gravity as far in front as possible. Thus, walking is a process of falling forward and then checking the fall. When we run, we lean farther forward, and so fall faster. ("Hygienic Physiology," p. 37.)

12. **The Pendulum** consists of a weight so suspended as to swing freely. Its movements to and fro are termed *vibrations* or *oscillations*. The path through which it passes is called the *arc*. The extent to which it goes in either direction from the lowest point is styled its *amplitude*. Vibrations performed in equal times are termed *i-soch'ronous* (*isos*, equal; *chronos*, time).

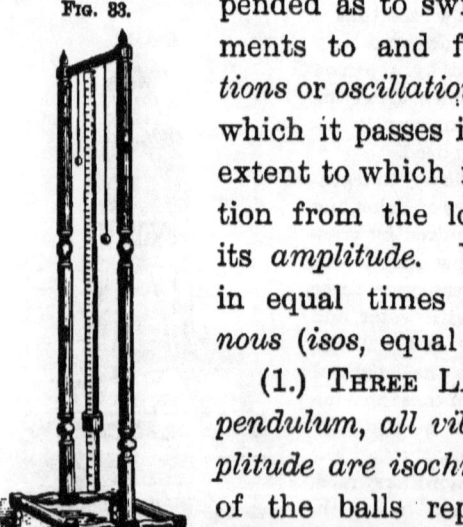

Fig. 33.

Pendulums.

(1.) THREE LAWS.—I. *In the same pendulum, all vibrations of small amplitude are isochronous.* If we let one of the balls represented in Fig. 33 swing through a short arc, and then through a longer one, on counting the number of oscillations per minute, we shall find them very uniform.

THE PENDULUM.

II. *The times of the vibrations of different pendulums are proportional to the square roots of their respective lengths.—Example:* A pendulum ⅑ the length of another, will vibrate three times as fast.* Conversely, the lengths of different pendulums are proportional to the squares of their times of vibration.

III. *The time of the vibration of the same pendulum will vary at different places,* since it decreases as the square root of the number expressing the acceleration of gravity increases. At the equator a pendulum vibrates most slowly. The length of a seconds-pendulum at New York is about $39\frac{1}{10}$ inches.

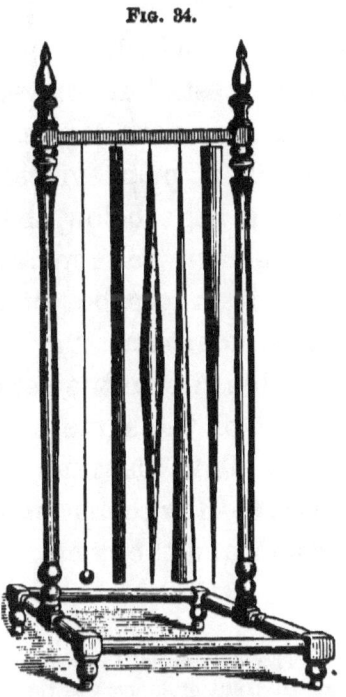

Fig. 34.

Pendulums of apparently the same length, but really different lengths.

(2.) CENTER OF OSCILLATION.—The upper part of a pendulum tends to move faster than the lower part,

---

* A pendulum which vibrates seconds must be four times as long as one which vibrates half-seconds. The apparatus represented in Figs. 33 and 34 can be made by any carpenter or ingenious pupil, and will serve excellently to illustrate the three laws of the pendulum. The law of the pendulum may be conveniently expressed in symbols. If t be the time of a single vibration in seconds, l the length of the pendulum, g the acceleration of gravity, l and g being expressed in feet, or in meters, and if π be the ratio of the circumference to the diameter of a circle, then

$$t = \pi\sqrt{\frac{l}{g}}.$$

This formula is convenient for use in solving problems.

and so hastens the speed. The lower part of a pendulum tends to move slower than the upper part, and so retards the speed. Between these extremes is a point which is neither quickened nor impeded by the rest, but moves in the same time that it would if it were a particle swinging by an imaginary line. This point is called the *center of oscillation*. It lies a little below the center of gravity.* In Fig. 34 is shown an apparatus containing pendulums of different shapes, but of the same length. If they are started together, they will immediately diverge, no two vibrating in the same time. As pendulums, they are not of the same length.

(3.) THE CENTER OF OSCILLATION IS FOUND BY TRIAL.†—Huyghens discovered that the point of suspension and the center of oscillation are interchangeable. If, therefore, a pendulum be inverted, and a

---

* This determines the real length of a pendulum, which is the distance from the point of support to the center of oscillation. The imaginary pendulum above described is known in Physics as the *Simple Pendulum*.—39.1 inches = 993.3 mm.

† "Take a flat board of any form and drive a piece of wire through it near its edge, and allow it to hang in a vertical plane, holding the ends of the wire by the finger and thumb. Take a small bullet, fasten it to the end of a thread, and allow the thread to pass over the wire so that the bullet hangs close to the board. Move the hand by which you hold the wire horizontally in the plane of the board, and observe whether the board moves forward or backward with respect to the bullet. If it moves forward, lengthen the string; if backward, shorten it till the bullet and the board move together. Now mark the point of the board opposite the center of the bullet, and fasten the string to the wire. You will find that, if you hold the wire by the ends and move it in any manner, however sudden and irregular, in the plane of the board, the bullet will never quit the marked spot on the board. Hence this spot is called the center of oscillation, because, when the board is oscillating about the wire when fixed, it oscillates as if it consisted of a single particle placed at the spot. It is also called the center of percussion, because, if the board is at rest and the wire is

THE PENDULUM. 71

point found at which it will vibrate in the same time as before, this is the former center of oscillation; while the old point of suspension becomes the new center of oscillation.*

(4.) THE PENDULUM AS A TIME-KEEPER.—The friction at the point of suspension, and the resistance of the air, soon destroy the motion of the pendulum. The clock is a machine for keeping up the vibration of the pendulum, and counting its beats. In Fig. 35, R is the scape-wheel driven by the force of the clock-weight or spring, and mn the escapement, moved by the forked arm, AB, so that only one cog of the wheel can pass at each double vibration of the pendulum. Thus the oscillations are counted by the cogs on the wheel, while the friction and the resistance of the air are overcome by the action of the weight or spring.† As " heat expands and cold contracts,"

Fig. 35.

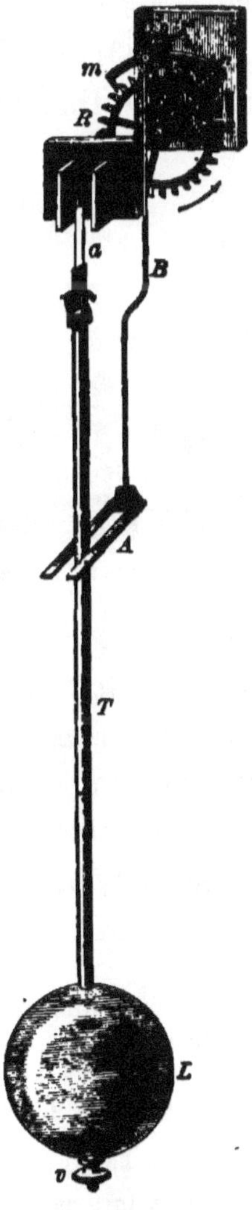

Clock Pendulum.

suddenly moved horizontally, the board will at first begin to rotate about the spot as a center."—J. CLERK MAXWELL, on "Matter and Motion," p. 104.

* The center of oscillation is the same as the *center of percussion*. The latter is the point where we must strike a suspended body, if we wish it to revolve about its axis without any strain. If we do not hit a ball on the bat's center of percussion, our hands " sting " with the jar.

† The action of a clock is clearly seen by procuring the works of an old clock and watching the movements of the various parts.

a pendulum lengthens in summer and shortens in winter. A clock, therefore, tends to lose time in summer and gain in winter. To regulate a clock, we raise or lower the pendulum-bob, *L*, by the nut *v*.

(5.) OTHER USES OF THE PENDULUM.—(*a.*) Since the time of vibration of a pendulum indicates the force of gravity, and the force of gravity decreases as the square of the distance from the center of the earth increases, we may thus find the semi-diameter of the earth at various places, and ascertain the figure of our globe. (*b.*) Knowing the force of gravity at any point, the velocity of a falling body can be determined. (*c.*) The pendulum may be used as a standard of measures. (*d.*) Foucault devised a method of showing the rotation of the earth on its axis, founded upon the fact that the pendulum vibrates constantly in one plane.* (*e.*) By observing the difference in the length of a seconds-pendulum at the

Fig. 36.

Foucault's Method.

---

* A pendulum 220 feet in length was suspended from the dome of the Pantheon in Paris. The lower end of the pendulum traced its vibrations north and south upon a table beneath, sprinkled with fine sand. These paths did not coincide, but at each return to the outside, the pendulum marked a point to the right. At the poles of the earth the pendulum, constantly vibrating in the same vertical plane, would perform a complete revolution in twenty-four hours, making thus a kind of clock. At the equator it would not change east or west, as the plane of vibration would go forward with the diurnal rotation of the earth. The shifting of the plane would increase as the pendulum was carried north or south from the equator.

top of a mountain and at the level of the sea, the density of the earth may be estimated.

PRACTICAL QUESTIONS.

1. When an apple falls to the ground, does the earth rise to meet it?
2. Will a body weigh more in a valley than on a mountain?
3. Will a pound weight fall more slowly than a two-pound weight?
4. How deep is a well if it takes three seconds for a stone to fall to the bottom?
5. Is the center of gravity always within a body—as, for example, a pair of tongs?
6. In a ball of equal density throughout, where is the center of gravity?
7. Why does a ball roll down hill?
8. Why is it easier to roll a round body than a square one?
9. Why is it easier to tip over a load of hay than one of stone?
10. Why is a pyramid such a stable structure?
11. When a hammer is thrown, on which end does it most often strike?
12. Why does a rope-walker carry a heavy balancing-pole?
13. What would become of a ball if dropped into a hole bored through the center of the earth?
14. Would a clock lose or gain time if carried to the top of a mountain? If carried to the North Pole?
15. In the winter, would you raise or lower the pendulum-bob of your clock?
16. Why is the pendulum-bob generally made flat?
17. What "beats-off" the time in a watch?
18. What should be the length of a pendulum to vibrate minutes at the latitude of New York? *Solution:* $(1 \text{ sec.})^2 : (60 \text{ sec.})^2 :: 39.1 \text{ in.} : x = 2.2 +$ miles.
19. What should be the length of the above to vibrate half-seconds? Quarter-seconds? Hours?
20. What is the proportionate time of vibration of two pendulums, respectively 16 and 64 inches long?
21. Why, when you are standing erect against a wall, and a piece of money is placed between your feet, can you not stoop forward and pick it up?
22. If a tower were 198 ft. high, with what velocity would a stone, dropped from the summit, strike the ground? (In these problems on falling bodies we may disregard the resistance of the air.)
23. A body falls in 5 seconds; with what velocity does it strike the ground?

ATTRACTION.

24. How far will a body fall in 10 seconds? With what velocity will it strike the ground?

25. A body is thrown upward with a velocity of 192 ft. the first second; to what height will it rise?

26. A ball is shot upward with a velocity of 256 ft.; to what height will it rise? How long will it continue to ascend?

27. Why do not drops of water, falling from the clouds, strike with a force equal to that calculated according to the laws of falling bodies? Because the mass of each drop is so small in proportion to its surface that the resistance of the air soon balances the acceleration of gravity, so that they fall with uniform velocity instead of accelerated velocity.

28. Are any two plumb-lines parallel?

29. A stone let fall from a bridge strikes the water in 3 seconds. What is the height?

30. A stone falls from a church-steeple in 4 seconds. What is the height of the steeple?

31. How far would a body fall in the first second at a distance of 12,000 miles above the earth's surface?

32. A body at the surface of the earth weighs 100 tons; what would be its weight 1,000 miles above?

33. A boy wishing to find the height of a steeple, lets fly an arrow that just reaches the top and then falls to the ground. It is in the air 6 seconds. Required the height.

34. An object let fall from a balloon reaches the ground in 10 seconds. Required the distance.

35. In what time will a pendulum 40 ft. long make a vibration?

36. Two bodies in space are 12 miles apart. Their masses are, respectively, 100 and 200 lbs. If they should fall together by their mutual attraction, what portion of the distance would be passed over by each body?

37. If a body weighs 2,000 lbs. upon the surface of the earth, what would it weigh 2,000 miles above? 500 miles above?

38. At what distance above the earth will a body fall, the first second, 21¼ inches?

39. How far will a body fall in 8 seconds? In the 8th second? In 10 seconds? In the 30th second?

40. How long would it take for a pendulum one mile in length to make a vibration?

41. What would be the time of vibration of a pendulum 64 meters long?

42. A ball is dropped from a height of 64 ft. At the same moment a second ball is thrown upward with sufficient velocity to reach the same point. How far from the ground will the two balls pass each other?

43. Explain the following fact: A straight stick loaded with lead at one end, can be more easily balanced vertically on the finger when the loaded end is upward than when it is downward.

44. If a body weighing a pound on the earth were carried to the sun it would weigh about 27 lbs. How much would it then attract the sun?

SUMMARY. 75

45. Why does watery vapor float and rain fall?

46. If a body weighs 10 kilos. on the surface of the earth, what would it weigh 1,000 kilometers above (the earth's radius being 6,366 km.)?

47. A body is thrown vertically upward with a velocity of 100 meters; how long before it will return to its original position?

48. Required the time needed for a body to fall a distance of 2,000 meters.

49. What would be the time of vibration of a pendulum 39.1 inches long at the surface of the moon, where the acceleration of gravity is only 4.8 ft.?

50. What would be the time of vibration for the same pendulum at the surface of the sun, where the acceleration of gravity is 27 times what it is at the earth's surface?

51. How many vibrations per minute would be made at the surface of the moon by a pendulum 40 ft. long?

52. A pendulum vibrates 200 times in 15 minutes. What is its length?

53. For a certain clock in New York the pendulum was made 500 lbs. in weight. What was the object in making it so heavy?

54. Pendulums are often supported by knife-edges of steel resting on plates of agate. Why?

55. The acceleration of gravity at the equator is 32.088 ft.; at the pole, 32.253 ft. If a pendulum vibrates 3,600 times an hour at the equator, how many times an hour will it vibrate at the pole?

SUMMARY.

THERE are certain forces operating between molecules and acting only at insensible distances, which are known as the Molecular Forces. The one which ties together molecules of the same kind is styled cohesion. The relation between this force and that of heat chiefly determines whether a body is solid, liquid, or gaseous. Under the action of cohesion, liquids tend to form spheres; and many solids, crystals. The processes of welding and tempering, and the annealing of iron and glass, illustrate curious modifications of the cohesive force. Molecules of different kinds are held together by adhesion. Its action is seen in the use of cement, paste, etc., in the solution of solids, in capillarity, diffusion of gases, and osmose.

Gravitation, though weak,* compared with cohesion, acts

---

* As the attraction of gravitation acts so commonly upon great masses of matter, we are apt to consider it a tremendous force. We, however, readily detect its relative feebleness when we compare the weight of bodies

universally. Its force is directly as the product of the attracting and attracted masses, and inversely as the square of their distance apart. Gravity makes a stone fall to the ground. The earth and a kilogram of iron in mid-air attract each other equally, but the mass of the former is so much greater that they move toward each other with unequal velocity, and the motion of the earth is imperceptible. Weight is the measure of the attraction of the earth. At the center of the earth the weight of a body would be nothing; at the poles it would be greatest, and at the equator least. Increase of distance above or far below the surface of the earth will diminish weight. Were the resistance of the air removed, all bodies would fall with equal rapidity. The laws of falling bodies may be studied with the aid of Atwood's Machine. The first second a body falls 16 ft. (4.9 meters), and gains a velocity of 32 ft. (9.8 meters). In general, the final velocity of a falling body is 32 ft., multiplied by the number corresponding to the second, and the distance is 16 ft. multiplied by the square of the number expressing the seconds. The center of gravity is the point about which the weights of all the particles composing a body will balance one another, $i. e.$, be in equilibrium. There are three states of equilibrium—stable, unstable, and indifferent—according as the point of support in a body is above, below, or at the center of gravity. As the center of gravity tends to seek the lowest point, its position determines the stability of a body. A body suspended so as to swing freely is a pendulum. The time of a pendulum's vibration is independent of its material, proportional to the square root of its length and variable according to the latitude. The pendulum is our time-keeper and useful in many scientific investigations.

We are so accustomed to see all the objects around us possess weight, that we can hardly conceive of a body deprived of a property which we are apt to consider as an essential attribute of matter. Nothing is more natural, apparently, than the falling of a stone to the ground. "Yet," says D'Alembert, "it is not without reason that philosophers are astonished to see a stone fall, and those who laugh at their astonishment would

---

with their tenacity.—*Example:* Think how much easier it is to lift an iron wire against gravity than to pull it to pieces against cohesion.

soon share it themselves, if they would reflect on the subject." Gravity is constantly at work about us, at one moment producing equilibrium or rest, and at another, motion. When it seems to be destroyed, it is only counterbalanced for a time, and remains, apparently, as indestructible as matter itself. The stability and the incessant changes of nature are alike due to its action. Not only do rivers flow, snows fall, tides rise, and mountains stand in obedience to gravitation, but smoke ascends and clouds float through the combined influence of heat and weight.

HISTORICAL SKETCH.

THE latter part of the sixteenth century witnessed the establishment of the principles of falling bodies. Galileo, while sitting in the cathedral at Pisa and watching the swinging of an immense chandelier which hung from its lofty ceiling, noticed that its vibrations were isochronous. This was the germ-thought of the pendulum and the clock. Up to his time it had been taught that a 4-lb. weight would fall twice as fast as a 2-lb. one. He proved the fallacy of this view by dropping from the Leaning Tower of Pisa balls of different metals—gold, copper, and lead. They all reached the ground at nearly the same moment. The slight variation he correctly accounted for by the resistance of the air, which was not the same for all.

Newton and his immediate predecessors knew the law of terrestrial gravity as manifested in falling bodies. When quite a young man, Newton entertained the idea that the attraction which draws bodies downward at the earth's surface must exist also between masses widely separated in space, such as the earth and the moon. To test this, he calculated how far the moon bends from a straight line, *i. e.*, falls toward the earth every second. Knowing the distance a body falls in a second at the surface of the earth, he endeavored to see how far it would fall at the distance of the moon. For years he toiled over this problem, but an erroneous estimate of the earth's diameter then accepted by physicists prevented his obtaining a correct result. Finally, a more accurate measurement having been made, he inserted this in his calculations. Finding the result was likely to verify his conjecture, his hand faltered with the excitement, and he was forced to ask a friend to complete the task. The truth was reached at last, and the grand law of gravitation discovered (1682).

The sun-dial was doubtless the earliest device for keeping time. The clepsydra was afterward employed. This consisted of a vessel containing water, which slowly escaped into a dish below, in which was a float that by its height indicated the lapse of time. King Alfred used candles of a uniform size, six of which lasted a day. The first clock erected in England, about 1288, was considered of so much importance that a high official was appointed to take charge of it. The clocks of the middle ages were extremely elaborate. They indicated the motions of the heavenly bodies; birds came out and sang songs, cocks crowed, and trumpeters blew their horns; chimes of bells were sounded, and processions of dignitaries and military officers, in fantastic dress, marched in front of the dial and gravely announced the time of day. Watches were made at Nuremberg in the fifteenth century. They were styled Nuremberg eggs. Many were as small as the watches of the present day, while others were as large as a dessert-plate. They had no minute or second hand, and required winding twice per day.

On Attraction, as well as on subsequent topics treated in this book, consult Guillemin's "Forces of Nature;" Atkinson's "Ganot's Physics"; Arnott's "Elements of Physics"; Snell's "Olmstead's Natural Philosophy"; Stewart's "Elementary Physics"; Silliman's "Physics"; Everett's "Text-book of Physics"; Young's "Lectures on Natural Philosophy"; "Appleton's Cyclopedia," articles on Clocks and Watches, Weights and Measures, Gravitation, Mechanics, etc.; Peck's "Ganot's Natural Philosophy"; Miller's "Chemical Physics," Chap. III., on Molecular Force; Weinhold's "Experimental Physics"; Pickering's "Elementary Physical Manipulation"; "Fourteen Weeks in Astronomy," sections on Galileo and Newton, pp. 29-34.

The current numbers of "Harper's Magazine," "The Century Magazine," "Scribner's Magazine," "Popular Science Monthly," "Boston Journal of Chemistry," "Scientific American," "Knowledge," and "Nature," contain the latest phases of science.

IV.
Elements of Machines.

Nature is a reservoir of power. Tremendous forces are all about us, but they are not adapted to our use. We need to remold the energy to fit our wants. A water-fall can not grind corn nor the wind draw water. Yet a machine will gather up these wasted forces, and turn a grist-mill or work a pump. A kettle of boiling water has little of promise; but husband its energy in the steam-engine, and it will weave cloth, forge an anchor, or bear our burdens along the iron track.

" The hero in the fairy tale had a servant who could eat granite rocks, another who could hear the grass grow, and a third who could run a hundred leagues in half an hour. So man in nature is surrounded by a gang of friendly giants who can accept harder stints than these. There is no porter like gravitation, who will bring down any weight you can not carry, and if he wants aid, knows how to get it from his fellow-laborers. Water sets his irresistible shoulder to your mill, or to your ship, or transports vast bowlders of rock, neatly packed in his iceberg, a thousand miles."

Emerson.

ANALYSIS OF THE ELEMENTS OF MACHINES.

ELEMENTS OF MACHINES.
- — The Simple Machines.
- — The Law of Mechanics.
- 1. The Lever.
  - 1. Definition.
  - 2. Three Classes of Levers.
    - (1.) *First Class.*
    - (2.) *Second Class.*
    - (3.) *Third Class.*
  - 3. Law of Equilibrium.
  - 4. Steelyard.
  - 5. Compound Lever.
- 2. The Wheel and Axle.
  - 1. Definition and Illustration.
  - 2. Law of Equilibrium.
  - 3. Wheel-work.
- 3. The Inclined Plane.
  - 1. Definition and Illustration.
  - 2. Law of Equilibrium.
- 4. The Screw.
  - 1. Definition and Illustration.
  - 2. Law of Equilibrium.
- 5. The Wedge.
  - 1. Definition and Illustration.
  - 2. Law of Equilibrium.
- 6. The Pulley.
  - 1. Definition and Illustration.
  - 2. Fixed and Movable Pulleys.
  - 3. Combinations of Pulleys.
  - 4. Law of Equilibrium.
- 7. Cumulative Contrivances.
- 8. Perpetual Motion.

ELEMENTS OF MACHINES.

The Simple Machines are the elements to which all machinery can be reduced. The watch with its complex system of wheel-work, and the engine with its belts, cranks, and pistons, are only various modifications of some of the six elementary forms—the *lever*, the *wheel and axle*, the *inclined plane*, the *screw*, the *wedge*, and the *pulley*. These six may be still further reduced to two—the lever and the inclined plane.

They are often termed the Mechanical Powers, but they do not produce work; they are only the means of applying it. Here again the doctrine of the Conservation of Energy holds good. The work done by the power is always equal to the resistance overcome in the weight.

The Law of Mechanics is, *the power multiplied by the distance through which it moves, is equal to the weight multiplied by the distance through which it moves.—Example:* 1 lb. of power moving through 10 feet = 10 lbs. of weight moving through one foot, or *vice versa*. In theory, the parts of a machine have no weight, move with no friction, and meet no resistance from the air. In practice, these influences must be considered.

ELEMENTS OF MACHINES.

1. The Lever is a bar turning on a pivot. The force used is termed the *power* (*P*), the object to be lifted the *weight* (*W*), the pivot on which the lever turns the *fulcrum* (*F*), and the parts of the lever each side of the fulcrum the *arms*.

THREE CLASSES OF LEVERS.—In the three kinds, the fulcrum, weight, and power are each respectively be-

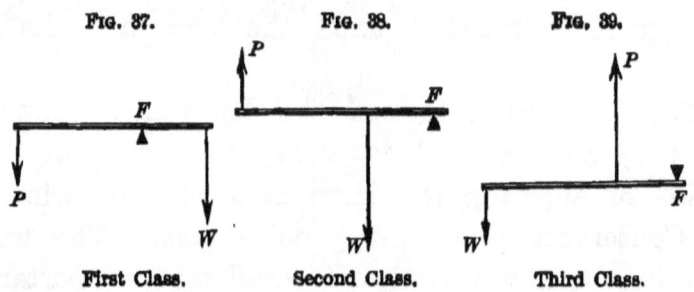

Fig. 37. First Class. Fig. 38. Second Class. Fig. 39. Third Class.

tween the other two, as may be seen by comparing Figs. 37-39.

First Class.—We wish to lift a heavy stone. Accordingly we put one end of a handspike under it, and resting the bar on a block at *F*, bear down at

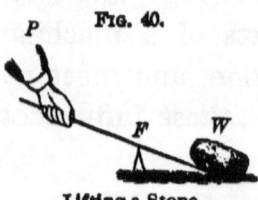

Fig. 40. Lifting a Stone.

P.—A pump-handle is a lever of the first class. The hand is the *P*, the water lifted the *W*, and the pivot the *F*.—A pair of scissors is a double lever of the same class. The cloth to be cut is the *W*, the hand the *P*, and the rivet the *F*.

Second Class.—We may also raise the stone by resting one end of the lever on the ground, which acts as a fulcrum, and lifting up on the bar.—An oar is a lever of the second class. The hand is the

P, the boat the W, and the water the F. In this case the F is not immovable.

Third Class.—The treadle of a sewing-machine is a lever of the third class. The front end resting on the ground is the F, the foot is the P, and the force is transmitted by a rod to the W, the arm above.

LAW OF EQUILIBRIUM.—The product of P multiplied by the perpendicular distance between its line of action and F, is called the *moment* of P. In the lever, P balances W when the moments about the fulcrum are equal. In Fig. 41, assume AB to be the initial position of a lever, which is then turned into the position $A'B'$ by application of the power, P, which balances the weight W, its line of action being $A'P$, while that of W is $B'W$. The power moves through a distance equal to AA', while the weight moves through a distance equal to BB'. But these distances are proportional to $A'F$ and $B'F$. We may represent $A'F$ by Pd, the distance of the power's line of action from F; and $B'F$ by Wd', the distance of the weight's line of action from F. Substituting these terms in the general expression of the law, we have,

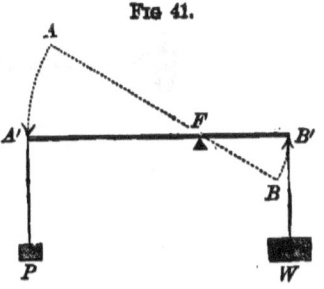

Fig 41.

$$P \times Pd = W \times Wd' \quad (1) \qquad P : W :: Wd' : Pd \quad (2) \qquad P = \frac{W \times Wd'}{Pd} \quad (3)$$

In the first and second classes, as ordinarily used, we gain power and lose time; in the third class we lose power and gain time.

84 ELEMENTS OF MACHINES.

The BALANCE is a lever of the first class with equal arms. The bar, AB (Fig. 42), has a pair of scale pans suspended from its ends. At the middle

FIG. 42.

The Balance.

an axis, n, made of steel and provided with a knife edge, rests upon a hard surface, so that the friction may be the least possible; this is the fulcrum.

The STEELYARD (Fig. 43) is a lever of the first class. The P is at E, the F at C, and the W at D. If the dis-

THE LEVER.

tance from the pivot of the hook *D* to the pivot of the hook *C* be one inch, and from the pivot of the hook *C* to the notch where *E* hangs be 12 inches, then 1 lb. at *E* will balance 12 lbs. at *W*. If the steelyard be reversed (Fig. 44), then the distance of the *F* from the *W* is only ¼ as great, and 1 lb. at *E* will balance 48 lbs. at *D*. Two sets of notches on opposite sides of the bar correspond to these different positions.

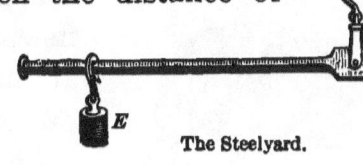

Fig. 43.

The Steelyard.

The COMPOUND LEVER consists of several levers so connected that the short arm of the first acts on the long arm of the second, and so on to the last of the series. If the distance of *A* (Fig. 45) from the *F* be four times that of *B*, a *P* of 5 lbs. at *A* will balance a *W* of 20 lbs. at *B*. If the arms of the second lever are of the same comparative length, a *P* of 20 lbs. at *C* will balance 80 lbs. at *E*. In the third lever, a *P* of 80 lbs. at *D* will balance 320 lbs. at *G*. With this system of three levers, 5 lbs. at *A* will accordingly balance 320 lbs. at *G*. To raise the *W* 1 ft., however, the *P* must move 64 ft. Thus what is gained in power is lost in time.

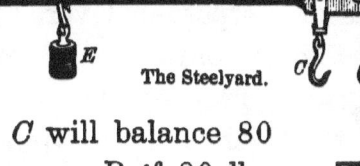

Fig. 44.

The Steelyard.

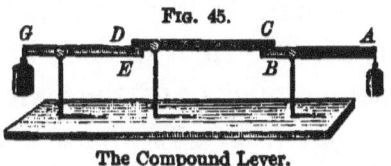

Fig. 45.

The Compound Lever.

There is no creation of force by the use of the levers; on the contrary, there is an appreciable loss because of friction.

Hay scales are constructed upon the principle of the compound lever. Considering the large mass on the platform as the power, its pressure is transmitted at the points P and P' (Fig. 46) to a pair of levers of the third class, whose fulcrums are at F and F'. Pressure is thus produced at P'' on another lever whose fulcrum is at F'''. At the remote end of this in turn, pressure is transmitted by the upright bar to the end, P''', of a lever of the first class whose fulcrum is at F''''. The weight, W, can be adjusted at will until a balance is secured.

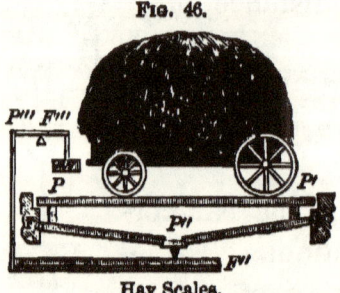
Fig. 46.
Hay Scales.

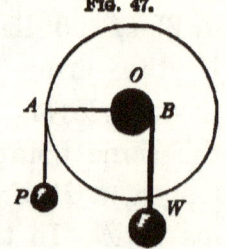
Fig. 47.
Wheel and Axle.

2. **The Wheel and Axle** is a kind of perpetual lever. As both arms work continuously, we are not obliged to prop up the W and re-adjust the lever. In the windlass used for drawing water from a well, the P is applied at the handle, the W is the bucket, and the F is the axis of the windlass. The long arm of the lever is the length of the handle, and the short arm is the semi-diameter of the axle. This is shown in a cross-section (Fig. 47) where the center, O, is the F, OA the long arm, and OB the short arm.—In Fig. 48, instead of turning a handle we take

THE WHEEL AND AXLE. 87

hold of pins inserted in the rim of the wheel.—Fig. 49 represents a capstan used on vessels for raising the anchor. The P is applied by handspikes inserted in the axle.—Fig. 50 shows a form of the capstan employed in moving buildings, in which a horse furnishes the power.

Fig. 48.

Wheel and Axle.

LAW OF EQUILIBRIUM.—By turning the handle or wheel around once, the rope will be wound around the axle and the W be lifted that distance. Applying the law of mechanics, $P \times$ the circumference of the wheel $= W \times$ circumference of the axle; or, as circles are proportional to their radii,

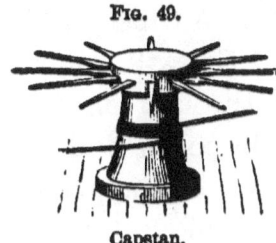

Fig. 49.

Capstan.

$P : W :: $ radius of the axle : radius of the wheel. . . . . (4)

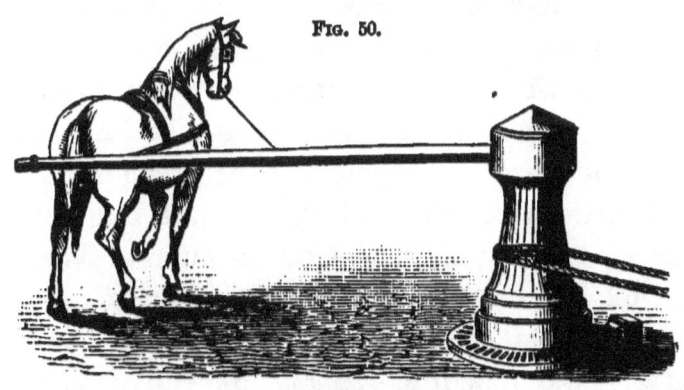

Fig. 50.

Capstan.

WHEEL-WORK consists of a series of wheels and axles which act upon one another on the principle of the compound lever. The projections on the circumference of the wheel are termed *teeth*, on the axle *leaves*, and the axle itself is called a *pinion*. If the radius of the wheel F be 12 inches, and that of each pinion 2 inches, then a P of 1 lb. will apply a force of 6 lbs. to the second wheel E. If the radius of this be 12 inches, then the second wheel will apply a P of 36 lbs. to the third wheel, which, acting on its axle, will balance a W of 216 lbs. The W will pass through only $\frac{1}{216}$ the distance of the P. We thus gain power and lose speed. If we wish to reverse this we can apply the P to the axle, and so gain speed. This is the plan adopted in factories, where a water-wheel furnishes abundant power, and spindles or other machines are to be turned with great rapidity.

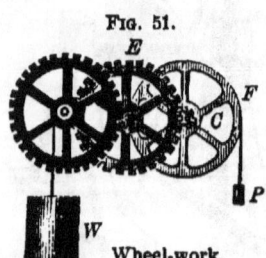

Fig. 51.

Wheel-work.

3. The Inclined Plane.—If we wish to lift a heavy cask into a wagon, we rest one end of a plank on the wagon-box and the other on the ground. We can then easily roll the cask up this inclined plane. When roads are to be made over steep hills, they are sometimes constructed around the hill, like the thread of a screw, or in a winding manner as shown in Fig. 52. There is a remarkable ascent of this kind on Mount Royal, Montreal.

LAW OF EQUILIBRIUM.—In Fig. 53 the P must descend vertically a distance equal to the length of the plane, AC, in order to move the W from A to C and thus elevate it through the vertical height BC. Ap-

FIG. 52.

Inclined Plane.

plying the law of mechanics, $P \times$ length of inclined plane $= W \times$ height of inclined plane; hence,

$P : W :: $ height of inclined plane : length of inclined plane.* (5)

If a road ascend 1 ft. in 100 ft., then a horse drawing up a wagon has to lift only $\frac{1}{100}$ of the load, besides overcoming the friction. A body sliding down a perfectly smooth inclined plane acquires the same velocity that it would in falling the same

FIG. 53.

Inclined Plane.

* If we roll into a wagon a barrel of pork, weighing 200 lbs., up a plane 12 ft. long and 3 ft. high, we have 12 ft. : 3 ft. :: 200 lbs. : $x =$ 50 lbs. We lift only 50 lbs., or ¼ of the barrel, but we move it through four times the space that would be necessary if we could elevate it directly into the wagon. We thus lose speed and gain power. The longer the inclined plane, the heavier the load we can lift, but the more time it will take to do it.

height perpendicularly. A train descending a grade of 1 ft. in 100 ft. tends to go down with a force equal to $\frac{1}{100}$ of its weight.*

4. The Screw consists of an inclined plane wound around a cylinder, the former being called the *thread*, and the latter the *body*. It works in a *nut* which is fitted with reverse threads to move on the thread of the screw. The nut may turn on the screw, or the screw in the nut. The P may be applied to either, by means of a wrench or lever. The screw is used in vises; in raising buildings; in copying letters, and in presses for squeezing the juice from apples, sugar-cane, etc.

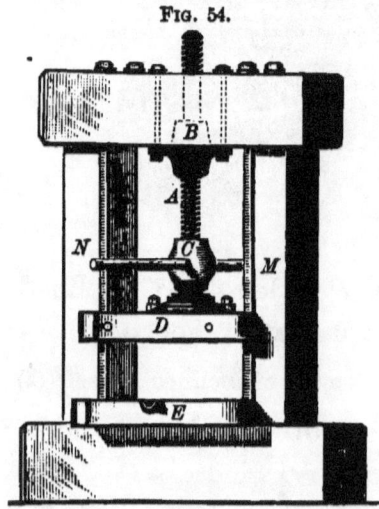

Fig. 54.

Screw.

LAW OF EQUILIBRIUM.— When the P is applied at the end of a lever, attached to the head of the screw, it describes a circle of which the lever is the radius. The distance through which the P passes, is the circumference of this circle; and the height to which the W is elevated at each revolution of the screw, is the distance between two of the threads.

---

* Near Lake Lucerne is a forest of firs on the top of Mount Pilatus, an almost inaccessible Alpine summit. By means of a wooden trough, the log is conducted into the water below, a distance of eight miles, in but little more than as many minutes. The force with which it falls is so prodigious, that if it jumps out of the trough it is dashed to pieces.

Applying the law of mechanics, $P \times$ circumference of circle $= W \times$ interval between the threads; hence,

$$P : W :: \text{interval} : \text{circumference.} \quad \ldots \ldots \quad (6)$$

The efficiency of the screw may be increased by lengthening the lever, or by diminishing the distance between the threads.

Fig. 55.

Wedge.

5. The Wedge consists generally of two inclined planes placed back to back. It is used for splitting wood and stone and lifting vessels in the dock. Leaning chimneys have been righted by wedges driven in on the lower side. Nails, needles, pins, knives, axes, etc., are made on the principle of the wedge.

THE LAW OF EQUILIBRIUM is the same as that of inclined plane—viz.:*

$$P : W :: \text{thickness of wedge} : \text{length of wedge.} \ldots (7)$$

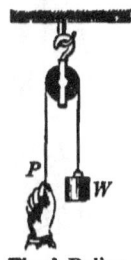

Fig. 56.

Fixed Pulley.

6. The Pulley consists of a wheel, within the grooved edge of which runs a cord.

A FIXED PULLEY (Fig. 56) merely changes the direction of the force. There is no gain of power or speed, as the hand P must move down as much as the weight W rises, and both with the same

---

* In practice, however, this by no means accounts for the prodigious power of the wedge. Friction, in the other mechanical powers, diminishes their efficiency; in this it is essential, else the wedge would fly back and the effect be lost. In the others, the P is applied as a steady pressure; in this it is a sudden blow, and depends upon the kinetic energy expended in the stroke of the hammer.

92　ELEMENTS OF MACHINES.

velocity. It is simply a lever of the first class with equal arms. By its use a man standing on the ground will hoist a flag to the top of a lofty pole, and thus avoid the trouble and danger of climbing up with it. Two fixed pulleys, arranged as shown in Fig. 57, make it possible to elevate a heavy load to the upper story of a building by horse-power.

FIG. 57. 
Application of Fixed Pulleys.

FIG. 58. 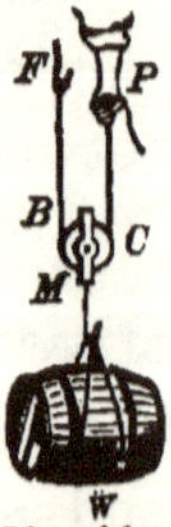
Movable Pulley.

A MOVABLE PULLEY is represented in Fig. 58. One half of the barrel is sustained by the hook while the hand lifts the other. As the P is one half the W, it must move through twice the space; in other words, by taking twice the time we can lift twice as much. Thus power is gained and time lost.

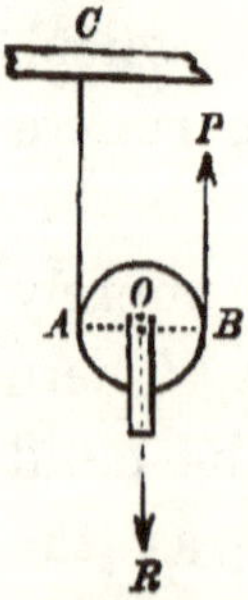

We may also explain the single movable pulley by Fig. 59. A represents the F, R the W acting in the line OR, and B the P acting in the line BP. This is a lever of the second class; and as $AO = \frac{1}{2}AB$, $P = \frac{1}{2}W$.

FIG. 59.
Movable Pulley.

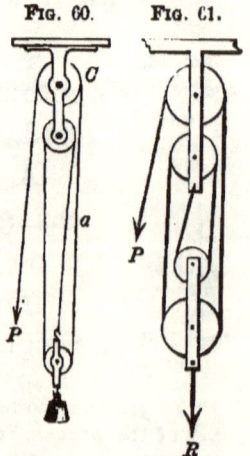

FIG. 60.　FIG. 61.
Systems of Pulleys.

COMBINATIONS OF PULLEYS. — (1.) In Fig. 60, we have the W sustained by three cords, each of which is stretched by a tension equal to the P; hence,

1 lb. of power will balance 3 lbs. of weight. (2.) In Fig. 61, the P will sustain a W of 4 lbs. (3.) In Fig. 62, the cord marked 1 1 has a tension equal to P in each part; the one marked 2 2 has a tension equal to $2P$ in each part, and so on with the others. The total tension acting on W is 16; hence, $W = 16P$. In this system, D rises twice as fast as C, four times as fast as B, etc. Work must stop when D reaches E, which gives little sweep to A for lifting W. (4.) Fig. 63 represents the ordinary "tackle-block" used by mechanics.

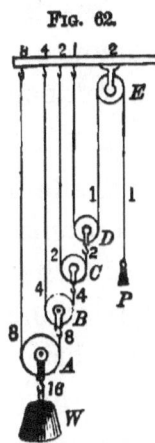

System of Pulleys.

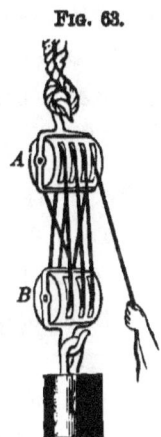

Tackle Block.

LAW OF EQUILIBRIUM.—When a continuous rope is used, let n represent the number of separate parts of the cord which sustain the movable block. We then have

$$P = \frac{W}{n}. \qquad \qquad (8)$$

When the number of movable and of fixed pulleys is equal, in general, $W = P \times$ twice the number of movable pulleys.

7. Cumulative Contrivances.— A hammer, club, pile-driver, sling, fly-wheel, etc., are instruments for accumulating energy to be used at the proper moment. Thus we may press a hammer on the head of a nail with all our strength to no purpose; but swing the hammer the length of the arm, and the blow will bury the nail to the head. The strength of our muscles and the attraction of gravity during the fall both gather energy to be exerted at the instant of contact. A fly-wheel by its momentum equalizes an irregular force, or produces a sudden effect.*

8. Perpetual Motion.—It is impossible to make a machine capable of perpetual motion. No combination can produce energy; it can only direct that which is applied. In all machinery there is friction; this must ultimately exhaust the power and bring the motion to rest. The only question is, how long a time will be required for the leakage to drain the reservoir. Every year brings to light new seekers after perpetual motion. The mere fact that a man devotes himself to the solution of this impossible problem is now generally regarded as a proof that either his mental balance has been disturbed, or his knowledge of the laws of nature is too meager to entitle him to consideration.

PRACTICAL QUESTIONS.

1. Describe the rudder of a boat as a lever. A door. A door-latch. A lemon-squeezer. A pitchfork. A spade. A shovel. A sheep-shears. A poker. A pair of tongs. A balance. A pair of pincers. A wheelbarrow. A man pushing open a gate with his hand near the hinge. A sledge-ham-

---

* We see the former illustrated in a sewing-machine, and the latter in a punch operated by a treadle. In the one case, the irregular action of the foot is turned into a uniform motion, and in the other it is concentrated in a heavy blow that will pierce a thick piece of metal.

mer, when the arm is swung from the shoulder. A nut-cracker. The arm (see "Physiology,' p. 48).

2. Show the change that occurs from the second to the third class of lever, when you take hold of a ladder at one end and raise it against a building.

3. Why is a pinch from the tongs near the hinge more severe than one near the end?

4. Two persons are carrying a weight of 250 lbs., hanging between them from a pole 10 ft. in length. Where should it be suspended so that one will lift only 50 lbs.?

5. In a lever of the first class, 6 ft. long, where should the F be placed so that a P of 1 lb. will balance a W of 23 lbs.?

6. What P would be required to lift a barrel of pork with a windlass whose axle is 1 ft. in diameter, and handle 3 ft. long?

7. What sized axle, with a wheel 6 ft. in diameter, would be required to balance a W of one ton by a P of 100 lbs.?

8. What number of movable pulleys would be required to lift a W of 200 lbs. by a P of 25 lbs.?

9. How many pounds could be lifted with a system of 4 movable pulleys by a P of 100 lbs.?

10. What W could be lifted with a single horse-power* acting on a system of pulleys shown in Fig. 62?

11. What distance should there be between the threads of a screw to enable a P of 25 lbs. acting on a handle 3 ft. long, to lift a ton?

12. How high would a P of 12 lbs., moving 16 ft. along an inclined plane, lift a W of 96 lbs.?

13. I wish to roll a barrel of flour into a wagon, the box of which is 4 ft. from the ground. I can lift but 24 lbs. How long a plank must I get?

14. What W can be lifted with a P of 100 lbs. acting on a screw having threads 1 in. apart, and a handle 4 ft. long?

15. What is the object of the balls often cast on the ends of the handle of the screw used in presses for copying letters?

16. In a steelyard 2 ft. long, the distance from the weight-hook to the fulcrum-hook is 2 in.; how heavy a body can be weighed with a pound weight?

17. Describe the change from the first to the third class of lever, in the different ways of using a pitchfork or a spade.

18. Why are not blacksmiths' tongs and fire-tongs constructed on the same principle?

19. In a lever of the third class, what W will a P of 50 lbs. balance, if one arm be 12 ft. and the other 3 ft. long?

---

* A *horse-power* is a force which is equivalent to 550 *foot-pounds*, *i. e.*, can raise against gravity 550 lbs. one foot in one second, or 33,000 lbs. one foot in one minute.

20. In a lever of the second class, what W will a P of 50 lbs. balance, with a lever 12 ft. long, and the W 3 ft. from the F?

21. In a lever of the first class, what W will a P of 50 lbs. balance, with a lever 12 ft. long, and the F 3 ft. from the W?

22. In a wheel and axle, the $P = 40$ lbs., the $W = 360$ lbs., and the diameter of the axle = 8 in. Required the circumference of the wheel.

23. Suppose, in a wheel and axle, the $P = 20$ lbs., the $W = 240$ lbs., and the diameter of the wheel = 4 ft. Required the circumference of the axle.

24. Required, in a wheel and axle, the diameter of the wheel, the diameter of the axle being 10 in., the P 100 lbs., and the W 1 ton.

25. Why is the rim of a fly-wheel made so heavy?

26. Describe the hammer, when used in drawing a nail, as a *bent lever*, *i.e.*, one in which the bar is not straight.

27. Describe the four levers shown in Fig. 46, when both the load of hay and the weight are considered, respectively, as the W and the P.

SUMMARY.

ALL machines can be resolved into a few elementary forms. Of these there are six, viz., the lever, the wheel and axle, the inclined plane, the screw, the wedge, and the pulley. Though called the mechanical powers, they are only instruments by which we can avail ourselves of the forces of nature. Molar energy or the motion of masses, as of air, water, steam, etc., is thus utilized, while the strength of a horse does the work of many men. A force of small intensity made to act through a considerable distance becomes one of great intensity acting through a small distance, and *vice versa*. By the use of the mechanical powers, the application of energy is made more convenient, but always some energy is absorbed in moving the machine and overcoming friction, and hence prevented from doing useful work. No machine can be a source of power, but, on the contrary, it thus involves a loss of power. The law of machines is, that the power multiplied by the distance through which it moves is equal to the weight multiplied by the distance through which it moves, plus the internal work involved in the motion of the machine. This law is equivalent to a statement that perpetual motion is impossible; for no known terrestrial source of energy is exhaustless.

The lever is a bar resting at some point on a prop as a center of motion. The crowbar, claw-hammer for drawing nails, pincers, windlass, and steelyard are examples of various classes of levers. The compound lever consists of several levers, so connected, that the short arm of one acts on the long arm of the next, as in the hay scales. In the bent lever, the power and weight act in lines that are not necessarily parallel, but still tend to produce rotation of the lever about its fulcrum, if the product of the power by the perpendicular distance from its line of action to the fulcrum be not equal to the weight multiplied by the distance from its line of action to the fulcrum. These two products are called the moments about the fulcrum. If the two moments are equal and opposite, the result is equilibrium.

To the lever may be reduced the wheel and axle, and the pulley. To the inclined plane may be reduced the wedge and the screw. The awl, vise, carpenter's plane, corkscrew, and stairs are common modifications of the inclined plane. The blade of a pocket-knife is a familiar example of the wedge, which itself is only a movable inclined plane. In the application of these last mechanical powers, friction becomes a most important and useful element; and it interferes so much with the operation of the simple machine alone, which should be devoid of friction in order to make exact calculation possible, that it is usually impossible to calculate the ratio between the power applied and the work accomplished through the medium of a wedge.

HISTORICAL SKETCH.

SIMPLE machines for moving large bodies are as old as history. The Babylonians knew the use of the lever, the pulley, and the roller. The Romans were acquainted with the lever, the wheel and axle, and the pulley (simple and compound). The Egyptians, it is thought, raised the immense stones used in building the Pyramids, by inclined planes made of earth which was afterward removed. Archimedes, in the third century B.C., discovered the law of equilibrium in the lever.* His investigations, however, were too far in advance of his time to be fully understood, and the teachings of Aristotle were long after accepted by scientific men. The law of mechanics, or of Virtual Velocities, as it is called, was discovered by Galileo.

* It is often said that Archimedes, in allusion to the tremendous power of the lever, asserted that, Give him a fulcrum and he could move the world. Had he been allowed such a chance, "the fulcrum being nine thousand leagues from the center of the earth, with a power of 200 lbs. the geometer would have required a lever 12 quadrillions of miles long and the power would have needed to move at the rate of a cannon-ball to lift the earth one inch in 27 trillions of years."

V.
PRESSURE OF LIQUIDS AND GASES.

"THE waves that moan along the shore,
The winds that sigh in blowing,
Are sent to teach a mystic lore
Which men are wise in knowing."

ANALYSIS.

PRESSURE OF LIQUIDS AND GASES.

I. HYDROSTATICS.

1. **LIQUIDS INFLUENCED BY EXTERNAL PRESSURE ONLY.**
    - (1.) Law of Transmission.
    - (2.) Water as a Mechanical Power.
    - (3.) Hydrostatic Press.

2. **LIQUIDS INFLUENCED BY GRAVITY.**
    - (1.) Four Laws of Equilibrium.
    - (2.) Rules for Computing Pressure.
    - (3.) Water Level.
    - (4.) Specific Gravity.
        - (a.) *Definition and Illustration.*
        - (b.) *Buoyant Force of Liquids.*
        - (c.) *To Find Specific Gravity of a Solid.*
        - (d.) *To Find Specific Gravity of a Liquid.*
        - (e.) *To Find Weight of Given Volume.*
        - (f.) *To Find Volume of Given Weight.*
        - (g.) *To Find Volume of a Body.*
        - (h.) *Floating Bodies.*

II. HYDRODYNAMICS.

—DEFINITION AND GENERAL PRINCIPLES.

1. **RULES CONCERNING A JET.**
    - (1.) The Velocity that of a Falling Body.
    - (2.) To Find the Velocity.
    - (3.) To Find the Quantity.
2. EFFECT OF TUBES.
3. FLOW OF WATER IN RIVERS.
4. **WATER-WHEELS.**
    - (1.) Overshot.
    - (2.) Undershot.
    - (3.) Breast.
    - (4.) Turbine.

III. PNEUMATICS.

—DEFINITION AND GENERAL PRINCIPLES.

1. AIR-PUMP.
2. CONDENSER.
3. **PROPERTIES OF AIR.**
    - (1.) Weight.
    - (2.) Elasticity.
    - (3.) Expansibility.
4. **PRESSURE OF THE AIR.**
    - (1.) The Proof.
    - (2.) Upward Pressure.
    - (3.) Buoyant Force.
    - (4.) Amount of Pressure.
    - (5.) Pressure Varies.
    - (6.) Mariotte's (or Boyle's) Law.
    - (7.) Barometer.
5. **PUMPS.**
    - (1.) Lifting.
    - (2.) Forcing.
    - (3.) Fire-engine.
6. SIPHON.
7. PNEUMATIC INKSTAND.
8. HYDRAULIC RAM.
9. ATOMIZER.
10. HEIGHT OF THE ATMOSPHERE.

Pressure of Liquids and Gases.

1. HYDROSTATICS.

HYDROSTATICS treats of liquids at rest. Its principles apply to all liquids; but water, on account of its abundance, is taken as the type.

1. Liquids Influenced by External Pressure only.
—(1.) PASCAL'S LAW.* *Liquids transmit pressure equally in all directions, and this acts at right angles upon the surface pressed.*

FIG. 64.

Illustration of Pascal's Law.

As the particles of a liquid move freely among themselves, there is no loss by friction, and any force will be transmitted equally upward, downward, and sidewise. Thus, if a bottle be filled with water and a pressure of 1 lb. be applied upon the cork, it will be communicated from particle to particle throughout the water. If the area of the cork be 1 sq. in., the pressure upon

---

* This law is named after the celebrated geometrician, Blaise Pascal, who first enunciated it in 1663. At first thought it may seem impossible for a pressure of 1 lb. to produce a pressure of 100 lbs.; but it should be remembered that the general law of mechanics applies to liquids as well as solids. If a force of 1 lb. on 1 sq. in. should *cause motion* by pressing through the medium of a liquid on 100 sq. in., the velocity of the body moved will be only $\frac{1}{100}$th of that of the body applying the pressure.

any sq. in. of the glass at *n, a, b,* or *c,* will be 1 lb. If the inside surface of the bottle be 100 sq. in., a pressure of 1 lb. upon the cork will produce a force of 100 lbs., tending to burst the bottle.

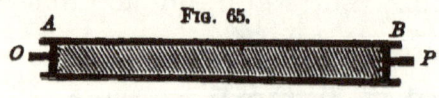

Fig. 65.
Tube with Cylinder of Lead.

Illustrations. — The transmission of pressure by liquids under some circumstances, is more perfect than by solids. Let a straight tube, *AB,* be filled with a cylinder of lead, and a piston be fitted to the end of the tube. If a force be applied at *P,* it will be transmitted to *O* without sensible loss. If instead, we use a bent tube, the force will be transmitted in the lines of the arrows, and will act on *P* but slightly. If, however, we fill the tube with water, the force will be transmitted without diminution.* Take a glass bulb and stem full of water, as in Fig. 67.† If you are careful to let the stem slip loosely through your fingers as the bulb strikes,

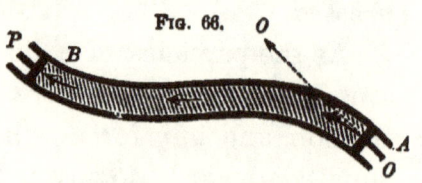

Fig. 66.
Bent Tube Full of Water.

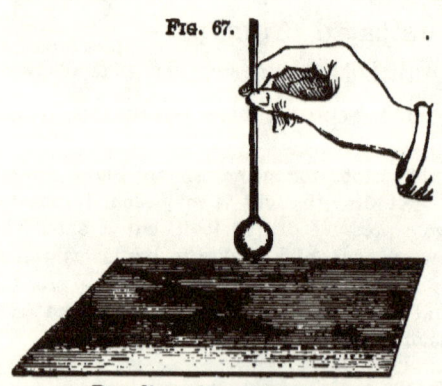

Fig. 67.
Pounding with a Glass Bulb.

---

* With cords, pulleys, levers, etc., we lose often more than one half of the force by friction; but this "liquid rope" transmits it with no appreciable loss.

† The process of filling such bulbs is shown on p. 248. They are cheaply

HYDROSTATICS.

you may pound it upon a smooth surface. The force of the blow is instantly transmitted from the thin glass to the water, which is almost incompressible, and this makes the bulb nearly as solid as a ball of glass. If a Rupert's drop be held in a closed vial of water so as not to touch the glass (Fig. 68), and the tapering end be broken, the water will transmit the concussion and shatter the vial.

Fig. 68.

Rupert's Drop in Vial.

(2.) WATER AS A MECHANICAL POWER.—Take two cylinders, P and p (Fig. 69), fitted with pistons and filled with water. Let the area of p be 1 sq. in. and that of P be 100 sq. in. Then a downward pressure of 1 lb. on each sq. in. of the small piston will produce an upward pressure of 1 lb. on each sq. in. of the large piston. Hence a P of 1 lb. moving at a rate of 1 in. per second will lift a W of 100 lbs. at a rate of $\frac{1}{100}$th of an inch per second.*

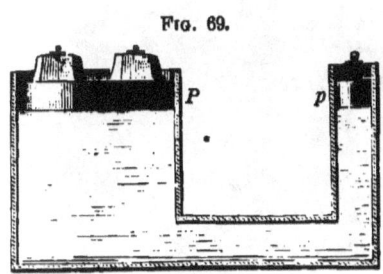

Fig. 69.

Principle of the Hydrostatic Press.

purchased of apparatus dealers, and explain not only this point, but also the method of filling thermometers.

* Pascal announced the discovery of this principle in the following words: "If a vessel full of water closed on all sides has two openings, the one a hundred times as large as the other, and if each be supplied with a piston which fits exactly, a man pushing the small piston will exert a force which will balance that of a hundred men pushing the large one."

104 PRESSURE OF LIQUIDS AND GASES.

(3.) THE HYDROSTATIC PRESS (Fig. 70) utilizes the principle just explained. As the workman depresses the lever O, he forces down the piston a upon the water in the cylinder A. The pressure is transmitted through the bent tube of water d under the piston

The Hydrostatic Press.

C, which lifts up the platform K, and compresses the bales. If the area of a be 1 in. and that of C 100 in., a force of 100 lbs. will balance 10,000 lbs. The handle is a lever of the second class. If the distance of the hand from the pivot be ten times that of the piston, a P of 100 lbs. will produce a force of 1,000 lbs. at a. This will become 100,000 lbs. at C.* According to the principle of mechanics, $P \times Pd = W \times Wd'$, the platform will ascend $\frac{1}{1000}$ of the distance the hand descends. This machine is used for baling hay and cotton, for launching vessels, and for testing the strength of ropes, chains, etc.

2. Liquids Influenced by Gravity.—The lower part of a vessel of water must bear the weight of the upper part. Thus each particle of water at rest is pressed downward by the weight of the minute column it sustains. It must, in turn, press in every direction with the same force, else it would be driven out of its place and the liquid would no longer be at rest. Indeed, when a liquid is disturbed it comes to rest—*i. e.*, there is an equilibrium established—only when this equality of pressure is produced. The following laws obtain:

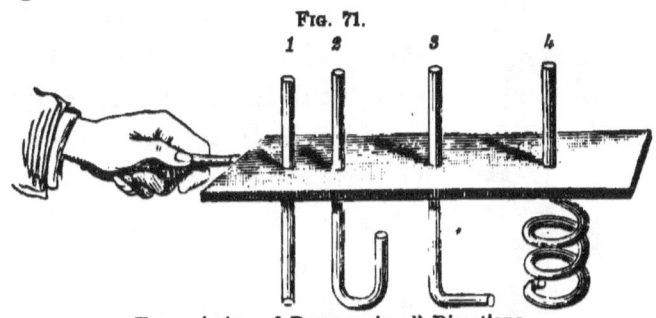

Transmission of Pressure in all Directions.

(1.) FOUR LAWS OF EQUILIBRIUM.—I. *At any point within a liquid at rest the pressure is the same in*

---

* The presses employed for raising the immense tubes of the Britannia Bridge across the Menai Strait, were each capable of lifting 2,672 tons, and of "throwing water in a vacuum to a height of nearly six miles, or over the top of the highest mountain."

106 PRESSURE OF LIQUIDS AND GASES.

all directions. If the series of glass tubes shown in Fig. 71 be placed in a pail of water, the liquid will be forced up 1 by the upward pressure of the water, 2 by the downward pressure, 3 by the lateral pressure, and 4 by the three combined in different portions of the tube. The water will rise in them all to the same height—*i. e.*, to the level of the water in the pail.

FIG. 72.

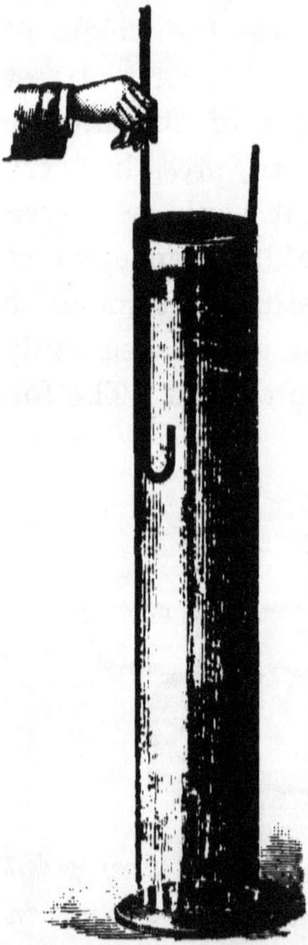

Increase of Pressure with Depth.

II. *The pressure increases with the depth.* Into a tall jar full of water put a bent tube open at both ends, as shown in Fig. 72, a little mercury having been previously poured in so as to fill the bend. The pressure of the water forces down the mercury in the short arm, that in the long arm being exposed only to the pressure of the air. Suppose the difference of level to be an inch when the tube is lowest. Then a column of mercury an inch long will just balance the weight of a column of water of the same thickness and nearly equal in length to the height of the jar. Let the tube now be raised until the surface of the mercury in the short arm is only a fourth of its previous distance from the surface. The difference of level in the two arms is now found to be only a fourth of an inch.

A cubic foot of water weighs about 62.5 lbs. (1,000 oz.); the same volume of sea-water weighs 64.4 lbs.; hence the pressure is proportionally greater in sea-water. At the greatest depth ever measured in the sea, a little over five miles, the pressure is about six tons on every square inch. An empty

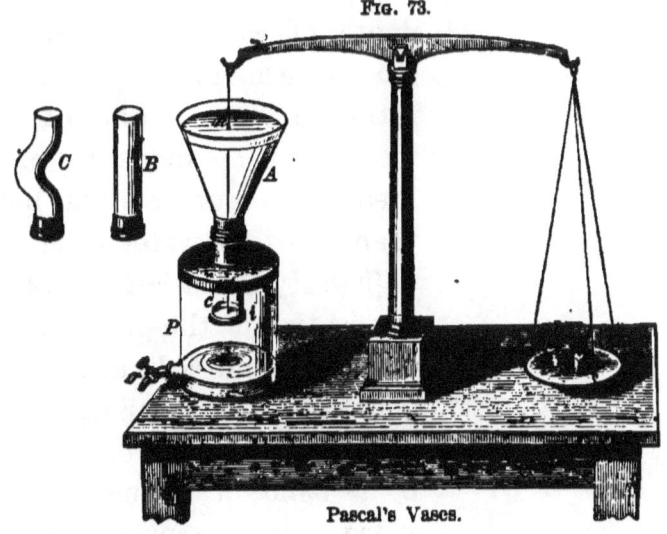

Fig. 73.

Pascal's Vases.

glass bottle securely stoppered may be crushed before sinking a hundred yards.

III. *The pressure does not depend on the shape or size of the vessel, but on the area and depth.* In the apparatus shown in Fig. 73, a disk is held up by a string against the bottom of an open tube, to which may be screwed vessels of different shapes and sizes, such as *A*, *B*, or *C*. The string is attached to the beam of a balance. In the scale pan is put such a weight, M, as to balance the pressure of the liquid against the disk when the vessel is full up to a certain point, n. The addition of any more liquid causes the disk to sink and thus spill the liquid, whether the large vessel, A, or the smaller ones, B or C, be used. Against the same area at the bottom the same pressure is obtained even if A is three times B in volume, provided the depth be kept the same. If the depth of water be increased, M has to be increased in the same ratio. Thus a pound of water in B may be made to exert a greater pressure at the bottom than 2 lbs. of water in A.

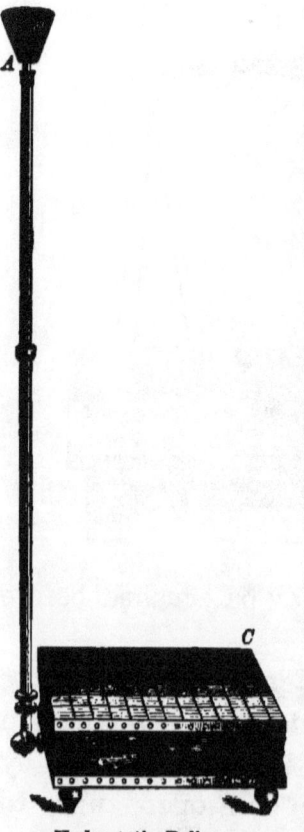

Fig. 74.

Hydrostatic Bellows.

The Hydrostatic Bellows is an application of the principles just discussed, and is like the Hydrostatic Press on a small scale. It consists of two boards connected by a band of leather and provided with a supply tube for water. Suppose the area of C (Fig. 74) to be 500 sq. in., and the area across the small tube A to be a single sq. in. Let the stiffness of the leather, along with an added weight on C, be equivalent to 100 lbs. Then only one fifth of a pound of water in A is needed to balance the 100 lbs. If at A we use a larger tube

HYDROSTATICS. 109

across which the area is 10 sq. in., then 2 lbs. of water in it are required to maintain equilibrium. The surface of the water in A will be about 6 in. above C, whether the large or small tube is used. If a cubic inch more be poured into the small tube, the same quantity will be forced into the bellows, so that C rises $\frac{1}{500}$th of an inch. If a larger weight is put on C, a higher column in A is required, but still C can be raised by any addition to A, however small.

A strong cask fitted with a small pipe 30 or 40 feet long, if filled with water will be burst asunder.* The pressure is as great as if the tube were of the same diameter as the cask. In a coffee-pot the small quantity of liquid in the spout balances the large quantity in the vessel. If it were not so, it would rise in the spout and run out.

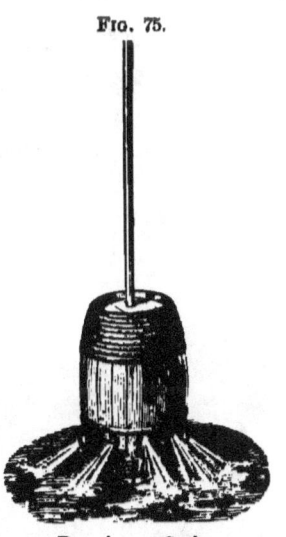

Fig. 75.

Bursting a Cask.

IV. *Water seeks its level.* In the apparatus shown in Fig. 76 the water rises to the same height in the various tubes which communicate with one another, because so long as the surfaces are not at the same level a particle below any surface

---

* Suppose the pipe to have an area of $\frac{1}{10}$ sq. in., and to hold $\frac{1}{4}$ lb. of water. The pressure on each $\frac{1}{10}$ of an inch of the interior of the cask would be $\frac{1}{4}$ lb., or 2,880 lbs. on each sq. ft.— a pressure no common barrel could sustain. The principle that a small quantity of water will thus balance another quantity, however large, or will lift any weight, however great, is frequently termed the "Hydrostatic Paradox." It is only an instance of a general law, and is no more paradoxical than the action of the lever.

110 PRESSURE OF LIQUIDS AND GASES.

must be unequally pressed from opposite sides, and must move until equilibrium is attained. In Fig. 77 a tank is situated on a hill, whence the water is conducted underground through a pipe to the fountain. In theory the jet will rise to the level of the reservoir, but in practice it falls short, owing chiefly to the friction in the pipe, the weight of the falling drops, and the resistance of the air.

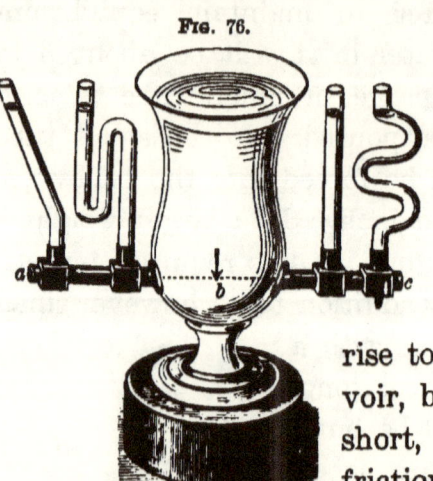

Fig. 76.

Water in Communicating Vessels.

Fig. 77.

Construction of a Fountain.

HYDROSTATICS. 111

Artesian Wells.* — Let A B and C D represent curved strata of clay impervious to water, and K K a layer of gravel. The rain falling on the hills filters down to C D, and collects in this basin. If a well be bored at H, as soon as it reaches the gravel the

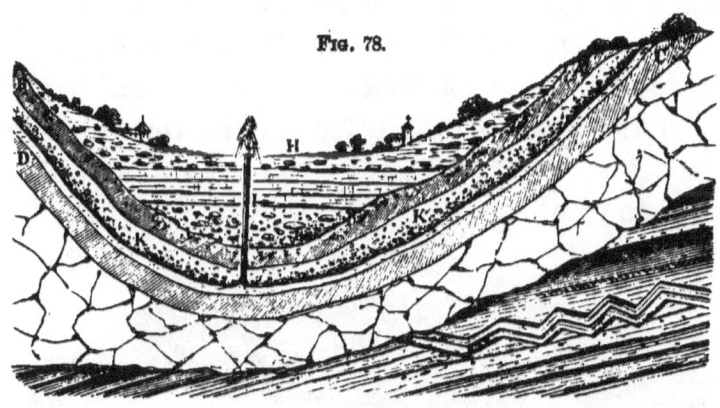

Fig. 78.

Artesian Well.

water will rush upward, under the tremendous lateral pressure, to the surface of the ground, often spouting high in air.†

Wells.—Of the rain which falls on the land, a part runs directly to the streams and part soaks into the soil. The latter portion may filter down to an im-

---

* They are so named because they have long been used in the province of Artois (Latin, Artésium), France. They were, however, early employed by the Chinese for the purpose of procuring gas and salt water.

† The famous well at Grenelle, Paris, is at the bottom of a basin which extends miles from the city. It is about 1,800 feet deep, and furnishes 656 gallons of water per minute. The two wells of Chicago are about 700 feet deep, and discharge daily about 432,000 gallons. Being situated on the level prairie, the force with which the water comes to the surface indicates that it may be supplied perhaps from Rock River, 100 miles distant. There are also valuable artesian wells at Louisville, Kentucky, and at Charleston, South Carolina. When the water comes from a great depth, it is generally warm.

112   PRESSURE OF LIQUIDS AND GASES.

permeable layer of rock or clay, and then run along till it oozes out at some lower point as a spring; or, if it can not escape, it will collect in the ground. If a well be sunk into this subterranean reservoir, the water will rise in it to the level of the source.*

(2.) RULES FOR COMPUTING PRESSURE.—I. *To find the pressure on the bottom of a vessel.* Multiply the area of the base in square feet by the vertical height in feet, and that product by the weight of a cubic foot of the liquid.

FIG 79.

The Curvature of a Water Level.

II. *To find the pressure on the side of a vessel.* Multiply the area of the side in square feet by half

---

* "From a forgetfulness of this principle the company which dug the Thames and Medway Canal, England, incurred heavy damages. Having planned the canal to be filled at high tide, the salt water spread immediately into all the wells of the surrounding region. Had the canal been dug a few feet lower, the evil would have been avoided."—ARNOTT.

of the vertical height in feet,* and that product by the weight of a cubic foot of the liquid. The pressure on the bottom of a cubical vessel of water is the weight of the water; on each side, one half; and on the four sides, twice the weight; therefore, on the five sides the pressure is three times the weight of the water.

(3.) WATER-LEVEL.—The surface of standing water is said to be level—*i. e.*, horizontal to a plumb-line. This is true for small sheets of water, but for larger bodies an allowance must be made for the spherical form of the earth (Fig. 79). The curvature is 8 inches for the first mile, and increases as the square of the distance.†

The *spirit-level* is an instrument used by builders for leveling. It consists of a slightly-curved glass tube nearly full of alcohol, so that it holds only a bubble of air. When the level is horizontal, the bubble remains at the center of the upper side of the tube.

Fig. 80.

The Spirit-level.

(4.) SPECIFIC GRAVITY, or relative weight, is the ratio of the weight of a substance to that of the

---

* This clause of the rule holds only when the center of gravity of the side is at half the vertical height. In general, the depth of its center of gravity below the surface should be used as the multiplier.

† For two miles it is 8 inches × 2^2 = 32 inches. If one's eye were at the level of still water, he could barely see the top of an object 67 feet high at a distance of 10 miles in a perfectly clear atmosphere.

same volume of another substance taken as a standard. Water is taken as the standard* for solids and liquids, and air for gases. A cubic inch of sulphur weighs twice as much as a cubic inch of water; hence its specific gravity = 2. A cubic inch of carbonic-acid gas weighs 1.52 times as much as the same volume of air; hence its specific gravity = 1.52.†

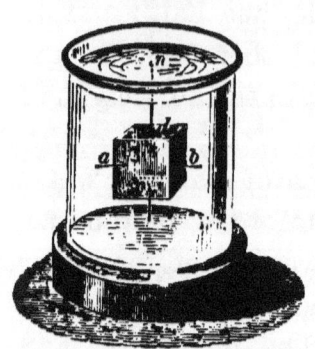

Fig. 81.

Immersed Cube.

Buoyant Force of Liquids.— The cube $a\ b\ c\ d$ is immersed in water. The lateral pressure at a is equal to that at b, because both sides are at the same depth; hence the body has no tendency toward either side of the jar. The upward pressure at c is greater than the downward pressure at d, because its depth is greater; hence the cube has a tendency to rise. This upward pressure is called the buoyant force of the water. *It is equal to the weight of the*

---

* "The water must be at 39.2° F., its greatest density. In all exact measurements, especially of standards, it is necessary to know the temperature. For the scale that is a foot long to-day may be more or less than a foot long to-morrow; the measure that holds a pint to-day may hold more or less than a pint to-morrow. Nay, more, these measures may not be the same in two consecutive moments. When a carpenter takes up his rule and applies it to some object, the size of which he wishes to determine, it becomes in that instant longer than it was before; when a druggist grasps his measuring glass in his hand to dispense some of his preparations, the glass increases in size. A person enters a cool room, and at once it becomes more capacious, for its walls, ceiling, and floor, because of the heat he imparts, immediately expand."—DRAPER.

† The term density is often used in the same sense as specific gravity, especially in relation to gases.

liquid displaced. For the downward pressure at d is the weight of a column of water whose area is that of the top of the cube, and whose vertical height is $n\ d$, and the upward pressure at c is equal to the weight of a column of the same size whose vertical height is $n\ c$. The difference between the two, or the buoyant force, is the weight of a volume of water equal to the size of the cube.

The same principle is shown in the "cylinder-and-bucket experiment." The cylinder a exactly fits in the bucket b. When the glass vessel in which the cylinder hangs is empty, the apparatus is balanced by weights placed in the scale-pan. Next, water is poured into the glass vessel. Its buoyant force raises the cylinder and depresses the opposite scale-pan. Then water is dropped into the bucket; when it is exactly full, the scales will balance again. This proves that "a body in water is buoyed up by a force equal to the weight of the water it displaces." This is called Archimedes' law.

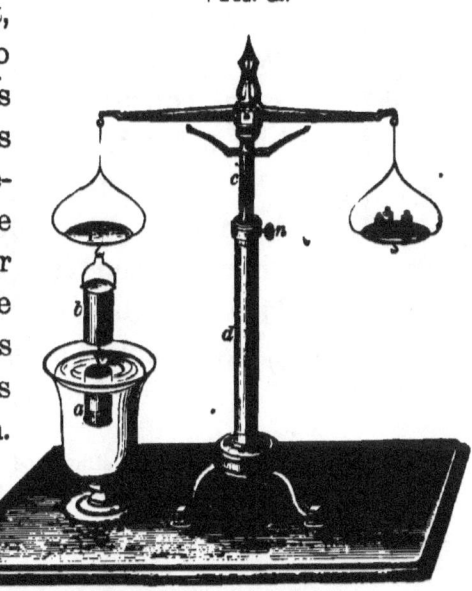

Fig. 82.

Cylinder-and-bucket Experiment.

116 PRESSURE OF LIQUIDS AND GASES.

To find the specific gravity of a heavy solid. Weigh the body in air, and in water; the difference is the weight of its volume of water; divide its weight in air by its loss of weight in water; the quotient is the specific gravity. Thus, sulphur loses one half its weight when immersed in water; hence it is twice as heavy as water, and its specific gravity $= 2$.*

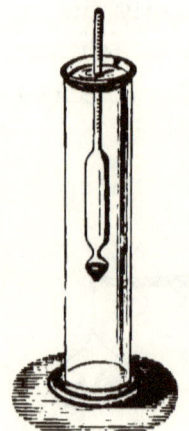

Fig. 83.

Hydrometer.

To find the specific gravity of a liquid by the specific-gravity flask. This is a bottle which holds exactly 1,000 grains of water. If it will hold 1,840 grains of sulphuric acid, the specific gravity of the acid is 1.84.

To find the specific gravity of a liquid by a hydrometer. This instrument consists of a glass tube, closed at one end and having at the other a bulb containing mercury. A graduated scale is marked upon the tube. The *alcoholometer*, used in testing alcohol, is so balanced as to sink in pure water to the zero point. As alcohol is lighter than water, the instrument will descend for every addition of spirits.

* In careful measurements an allowance is made for the weight of the air displaced by the body, so that its weight in a vacuum becomes known. Strictly, it is the weight in a vacuum that has to be compared with the loss of weight in water. If the body will not sink in water, attach it to a heavy body. 1. Weigh the lighter body in air (A). 2. Weigh the heavy body in water (B). 3. Weigh both together in water (C). Now C is less than B because the light body buoys up the heavy one; *i. e.*, its weight A is more than balanced, and is replaced by an upward or lifting force $= B - C$. Therefore the loss of the light body in water $= A + B - C$ $\therefore$ spec. grav. $= \dfrac{A}{A + B - C}$.

The degrees of the scale indicate the percentage of alcohol. Similar instruments are used for determining the strength of milk, acids, etc.

To find the weight of a given volume of any substance. Multiply the weight of one cubic foot of water by the specific gravity of the substance, and that product by the number of cubic feet.—*Example:* What is the weight of three cubic feet of cork? *Solution:* 1,000 oz. $\times$ .240* = 240 oz.; 240 oz. $\times$ 3 = 720 oz.

To find the volume of a given weight of any substance. Multiply the weight of a cubic foot of water by the specific gravity of the substance, and divide the given weight by that product. The quotient is the required volume in cubic feet.—*Example:* What is the volume of 20,000 oz. of lead? *Solution:* 1,000 oz. $\times$ 11.36 = 11,360; 20,000 $\div$ 11,360 = 1.76 + cu. ft.

To find the volume of a body. Weigh it in water. The loss of weight is the weight of the displaced water. Then, as a cubic foot of water weighs 1,000 oz., we can easily find the volume of water displaced. — *Example:* A body loses 10 oz. on being weighed in water. The displaced water weighs 10 oz. and is $\frac{1}{100}$ of a cubic foot; this is the exact volume of the body.

* TABLE OF SPECIFIC GRAVITY. (See "Chemistry," p. 288.)

Iridium... 21.80	Zinc ... 7.15	Pine Wood... .66
Platinum... 21.53	Diamond....about 3.50	Cork ... .24
Gold ... 19.34	Flint Glass... 2.76	Sulphuric Acid... 1.84
Mercury... 13.59	Chalk... 2.65	Water from Dead Sea. 1.24
Lead ... 11.36	Sulphur... 2.00	Milk ... 1.03
Silver ... 10.50	Ice ... .93	Sea-water... 1.03
Copper... 8.90	Potassium... .86	Absolute Alcohol ... .79
Cast-iron ... 7.21	Quicklime... .80	

Floating Bodies.—A body will float in water when its weight is not greater than that of an equal volume of the liquid, and its weight always equals that of the fluid displaced. An egg dropped into a glass jar half full of water (Fig. 84) sinks directly to the bottom. If, by means of a funnel with a long tube, we pour brine beneath the water, the egg will rise. We may vary the experiment by not dropping in the egg until we have half filled the jar with the brine. The egg will then fall to the center, and there float. Almost any solid, if dissolved in water, fills the pores of the water without adding much to its volume. This increases its density and buoyant power. A person can therefore swim more easily in salt than in fresh water.*—An iron ship will not only float itself, but also carry a heavy cargo, because it displaces a great volume of water.—A body floating in water has its center of gravity at the lowest point, when it is in stable equilibrium.†—Fishes have air-bladders, by which they can rise or sink

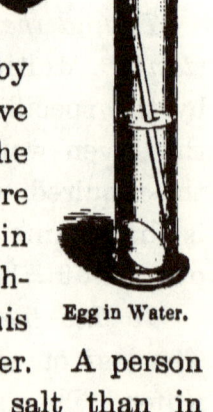

Fig. 84.

Egg in Water.

---

* Bayard Taylor says that he could float on the surface of the Dead Sea, with a log of wood for a pillow, as comfortably as if lying on a spring mattress. Another traveler remarks, that on plunging in he was thrown out again like a cork; and that on emerging and drying himself, the crystals of salt which covered his body made him resemble an "animated stick of rock-candy."

† Herschel tells an amusing story of a man who attempted to walk on water by means of large cork boots. Scarcely, however, had he ventured

at pleasure.* By compressing the air-bladder, the fish diminishes the volume of its own body. The buoyant effect of the water is correspondingly decreased and the fish descends. By relaxing the compression on the bladder, the air in it expands and the fish rises.

PRACTICAL QUESTIONS.

1. Why can housekeepers test the strength of lye by trying whether or not an egg will float on it?
2. How much water will it take to make a gallon of strong brine?
3. Why ought a fat man to swim more easily than a lean one?
4. Why does the firing of a cannon sometimes bring to the surface the body of a drowned person? *Ans.* Because by the concussion it shakes the body loose from the mud or any object with which it is entangled.
5. Why does the body of a drowned person generally come to the surface of the water after a time? *Ans.* Because the gases which are generated by decomposition in the body makes its specific gravity less.
6. If we let bubbles of air pass up through a glass of water, why will they become larger as they ascend?
7. What is the pressure on a canal lock-gate 14 feet high and 10 feet wide, when the lock is full of water?
8. Will a pail of water weigh any more with a live fish in it than without?
9. If the water filtering down through a rock should collect in a crevice an inch square and 250 feet high, opening at the bottom into a closed fissure having 20 square feet of surface, what would be the total pressure tending to burst the rock?
10. Why can stones in water be moved so much more easily than on land?
11. Why is it so difficult to wade in water when there is any current?

---

out ere the law of gravitation seized him, and all that could be seen was a pair of heels, whose movements manifested a great state of uneasiness in the human appendage below.

* It was formerly thought that a fish in water has no weight. It is said that Charles II. of England once asked the philosophers of his time to explain this phenomenon. They offered many wise conjectures, but no one thought of trying the experiment. At last a simple-minded man balanced a vessel of water, and on adding a fish, found it weighed just as much as if placed on a dry scale-pan.

120   PRESSURE OF LIQUIDS AND GASES.

12. Why is a mill-dam or canal embankment small at the top and large at the bottom?

13. In digging canals, ought the engineer to take into consideration the curvature of the earth?

14. Why does the bubble of air in a spirit-level move as the instrument is turned?

15. Can a swimmer tread on pieces of glass at the bottom of the water with less danger than on land?

16. Will a vessel displace more water in a fresh river than in the ocean?

17. Will iron sink in mercury?

18. The water in the reservoir in New York is about 80 feet above the fountain in the City Hall Park. What is the pressure upon a single inch of the pipe at the latter point?

19. Why does cream rise on milk?

20. There is a story told of a Chinese boy who accidentally dropped his ball into a deep hole where he could not reach it. He filled the hole with water, but the ball would not quite float. He finally thought of a successful expedient. Can you guess it?

21. Which has the greater buoyant force, water or oil?

22. What is the weight of four cubic feet of cork?

23. How many ounces of iron will a cubic foot of cork float in water?

24. What is the specific gravity of a body whose weight in air is 30 grs. and in water 20 grs.? How much is it heavier than water?

25. Which is heavier, a gallon of fresh or one of salt water?

26. The weights of a piece of syenite-rock in air and water were 3041.8 grs. and 2607.5 grs. Find its specific gravity.

27. A specimen of green sapphire from Siam weighed in air 21.45 grs. and in water 16.33 grs.; required its specific gravity.

28. A specimen of granite weighs in air 534.8 grs., and in water 334.6 grs.; what is its specific gravity?

29. What is the volume of a ton of iron? A ton of gold? A ton of copper?

30. What is the weight of a cube of gold 4 feet on each side?

31. A cistern is 12 feet long, 6 feet wide, and 10 feet deep; when full of water, what is the pressure on each side?

32. Why does a dead fish always float on its back?

33. A given volume of water weighs 62.5 grs., and the same volume of muriatic acid 75 grs. What is the specific gravity of the acid?

34. A vessel holds 10 lbs. of water; how much mercury would it contain?

35. A stone weighs 70 lbs. in air and 50 in water; what is its volume?

36. A hollow ball of iron weighs 10 lbs.; what must be its volume to float in water?

37. Suppose that Hiero's crown was an alloy of silver and gold, and weighed 22 oz. in air and 20½ oz. in water. What was the proportion of each metal?

38. Why will oil, which floats on water, sink in alcohol?

39. A specific gravity bottle holds 100 gms. of water and 180 gms. of sulphuric acid. Required the density of the acid.

40. What is the density of a body which weighs 58 gms. in air and 46 gms. in water?

41. What is the density of a body which weighs 63 gms. in air and 35 gms. in a liquid of a density of .85?

II. HYDRODYNAMICS.

HYDRODYNAMICS treats of liquids in motion. In this, as in Hydrostatics, water is taken as the type. In theory, its principles are those of falling bodies, but in practice they can not be relied upon except when verified by experiment. The discrepancy arises from changes of temperature which vary the fluidity of the liquid, from friction, the shape of the orifice, etc.

1. Rules Concerning a Jet.—(1.) THE VELOCITY OF A JET IS THE SAME AS THAT OF A BODY FALLING FROM THE SURFACE OF THE WATER. We can see that this must be so, if we recall two principles: First, "a jet will rise to the level of its source;" and second, "to elevate a body to any height, it must have the same velocity that it would acquire in falling that distance." It follows that the velocity of a jet depends on the height of the liquid above the orifice.

(2.) To FIND THE VELOCITY OF A JET OF WATER, use the 4th equation of falling bodies, $v^2 = 2gh$, in which h is the distance of the orifice below the surface of the water.—*Example:* The depth of water above the orifice is 49 feet; required the velocity Substituting, $v = \sqrt{2 \times 32 \times 49} = 56$ feet.

(3.) To FIND THE QUANTITY OF WATER DISCHARGED IN A GIVEN TIME, multiply the area of the orifice by the velocity of the water, and that product by the number of seconds.—*Example:* What quantity of water will be discharged in 5 seconds from an orifice having an area of $\frac{1}{2}$ sq. foot at an average depth of 49 feet? At that depth, $v = \sqrt{2 \times 32 \times 49} = 56$ feet per second; multiplying by $\frac{1}{2}$, we have 28 cubic feet discharged in one second and 140 cubic feet in five seconds.* In practice, much less than this can be realized.

2. Effect of Tubes.—If we examine a jet of water, we see its size is decreased just outside the orifice to about two thirds that at the opening. This neck is called the *vena contracta*, and is caused by the water producing cross currents as it flows from different directions toward the orifice. If a tube of a length twice or thrice the diameter of the opening be inserted, the water will adhere to the sides so that there will be no contraction, and the flow be increased to about 80 per cent. of the theoretical amount. If the tube be conical, and inserted with the large end inward, the discharge may be augmented to 95 per cent.; and if the outer end be

---

* If, at a foot below the surface, an opening will furnish 1 gallon per minute, to double that quantity the opening must be 4 feet below the top. Again, if a certain power will force through a nozzle of a fire-engine a given quantity of water in a minute, to double the quantity the power must be quadrupled.

flaring, it may reach 98 per cent. Long tubes or short angles, by friction, diminish the flow of water.

3. Flow of Water in Rivers.—A fall of three inches per mile is sufficient to give motion to water, and produce a velocity of as many miles per hour. The Ganges descends but 800 feet in 1,800 miles. Its waters require a month to move down this long inclined plane.* A fall of three feet per mile will make a mountain torrent. The current moves more swiftly at the center than near the shores or bottom of a channel, since there is less friction.

4. Water-wheels are machines for using the force of falling water. By bands or cog-wheels the motion of the wheel is conducted from the axle into the mill.†

Fig 85.

The Overshot Wheel.

The OVERSHOT WHEEL has on its circumference a series of buckets which receive the water flowing from a *sluice*, *C*. These hold the water as they descend on one side, and empty it as they come up on the other. Overshot wheels are valuable where a great fall can be secured, since they require but little water. If *W* denotes the weight of the water and *h* the distance it

---

* "The fall of 800 feet would theoretically give a velocity of more than 150 miles per hour. This is reduced by friction to about three miles."

† The principle is that of a lever with the *P* acting on the short arm. In this way the movement of the slow creaking axle reappears in the swiftly buzzing saw or flying spindle.

124   PRESSURE OF LIQUIDS AND GASES.

falls, then the total work $= Wh$. Of this amount, 75 per cent. can be made available under good conditions.

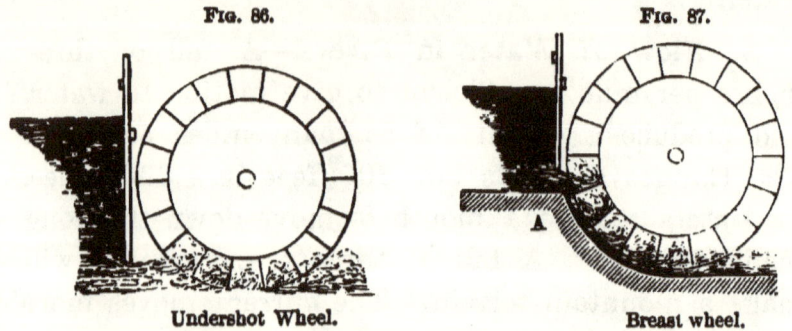

Fig. 86. Undershot Wheel.

Fig. 87. Breast wheel.

The UNDERSHOT WHEEL has projecting boards, or *floats*, which receive the force of the current. It is of use where there is little fall and a large quantity of water. It utilizes not more than 25 per cent. of the energy of the water.

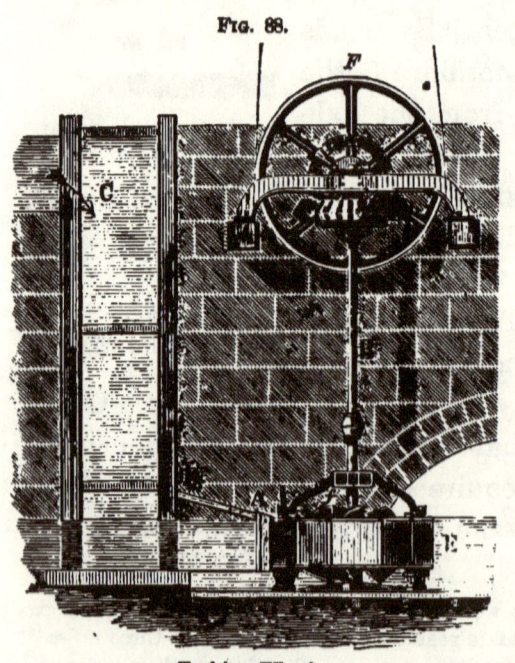

Fig. 88. Turbine Wheel.

The BREAST-WHEEL (Fig. 87) is a medium between the two kinds already named.

The TURBINE WHEEL is placed horizontally and immersed in the water. In Fig. 88, C is the dam and DA the spout

HYDRODYNAMICS. 125

by which the water is furnished. *E* is a scroll-like casing encircling the wheel, and open at the center above and below. The axis of the wheel is the vertical cylinder *B*, from which radiate plane-floats against which the water strikes. This form utilizes as high as 90 per cent. of the energy. *F* is a band-wheel which conducts the power to the machinery.

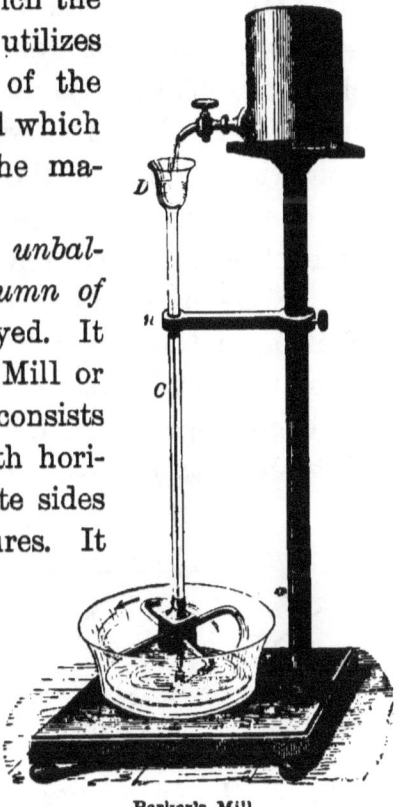

Fig. 80.

Barker's Mill.

The principle of the *unbalanced pressure of a column of water* may also be employed. It is illustrated in Barker's Mill or Reaction-wheel.* This consists of an upright cylinder with horizontal arms, on the opposite sides of which are small apertures. It rests in a socket, so as to revolve freely. Water is supplied from a tank above. If the openings in the arms are closed, when the cylinder is filled with water the pressure is equal in all directions and the machine is at rest. If now we open an aperture, the pressure is relieved

---

* Revolving fire-works and the whirligig, used for watering lawns and as an ornament in fountains, are constructed on the same principle.—An ingenious pupil can easily construct a Reaction-wheel of straws or quills, pouring the water into the upright tube by means of a pitcher, or admitting it slowly through a siphon from a pail of water placed on a table above.

on that side, and the arm flies back on account of the unbalanced pressure of the column of water above.

5. Waves are produced by the friction of the wind against the surface of the water. The wind raises the particles of water and gravity draws them back again. They thus vibrate up and down, but in deep water the liquid mass does not advance. The forward movement of the wave is an illusion. The *form* of the wave progresses like the apparent motion of the thread of the screw which we turn in our hand, or the undulations of a rope or carpet which is shaken, or the stalks of grain which bend in billows as the wind sweeps over them.

The corresponding parts of different waves are said to be *like phases*. Thus, in Fig. 90, A and E,

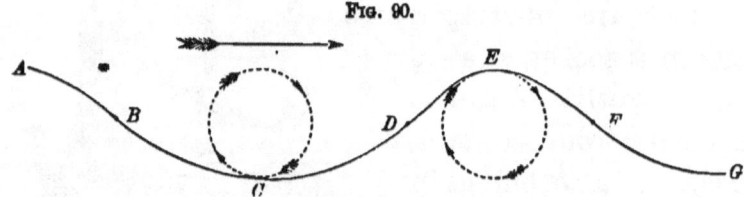

Fig. 90.

B and F, C and G are like phases. The distance between two like phases, or between the crests of two succeeding waves, is called a *wave-length*. Thus the distance AE, or BF, or CG is a wave-length. Opposite phases are those parts which are vibrating in opposite directions, as E and C, or B and D. The successive particles of water move each in an ellipse, and in regular succession, so that when a particle at

HYDRODYNAMICS. 127

E is moving forward, one at *C* is moving backward, one at *B* upward, and one at *D* downward. This is easily observed at sea.*

COMPOSITION OF WAVE MOTION.—A tide-wave may be setting steadily toward the west; waves from

FIG. 91.

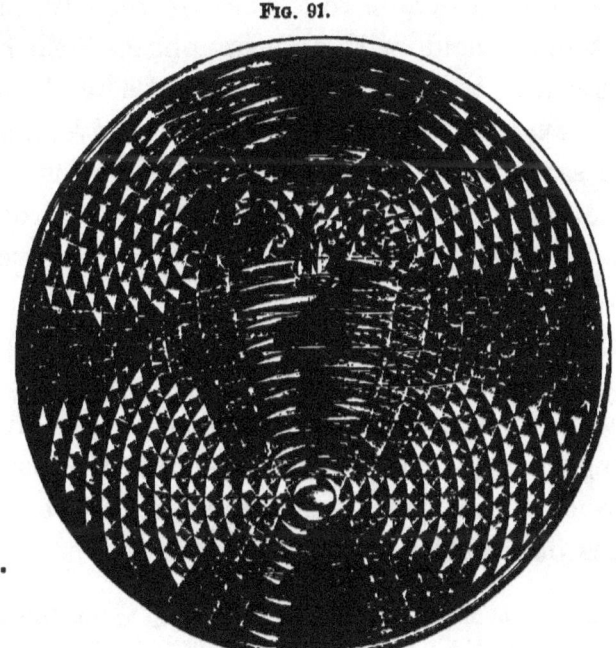

Interference of Waves.

distant storms may be moving upon this; and, above the whole, ripples from the breeze then blowing may diversify the surface. These different systems of

---

* Near the shore the oscillations become shorter; the lower particles being checked in their elliptic motion by the friction on the sandy beach, the front becomes well-nigh vertical, and the upper part curls over and falls beyond. The size of "mountain billows" has been exaggerated. Along the coast in a gorge they may reach 90 feet, but in the open sea the highest wave, from the deepest "trough" to the very topmost "crest," rarely measures over 30 feet.

waves will compound into a resultant system in accordance with the following general principles: If any two systems coincide with *like* phases,—the crest of one meeting the crest of the other, and the furrow of one meeting the furrow of the other,—the resulting wave will have a height equal to the *sum of the two*. If any two systems coincide with opposite phases,—the hollow of one striking the crest of another,—the height will be the *difference of the two*. Thus, if in two systems having the same wave-length and height, one is exactly half a length behind the other, they will destroy each other. This is termed the *interference* of waves.*

The manner in which different waves move among and upon one another, is seen by dropping a handful of stones in water and watching the waves as they circle out from the various centers in ever-widening curves. In Fig. 91 is shown the beautiful appearance these waves present when reflected from the sides of a vessel.

---

* "In the port of Batsha the tidal-wave comes up by two distinct channels so unequal in length that their time of arrival varies by six hours. Consequently when the crest of high water reaches the harbor by one channel, it meets the low water returning by the other, and when these opposite phases are equal, they neutralize each other, so that at particular seasons there is no tide in the port, and at other times there is but one tide per day, and that equal to the difference between the ordinary morning and evening tide."—*Lloyd's Wave Theory*.

Another striking example of interference of tide-waves is seen in the immediate neighborhood of New York. The tide-wave from the ocean, coming from the south-east, divides, a part passing up New York Bay, and another part sweeping around and turning westward through Long Island Sound. The meeting-place of these two branches is at Hell Gate, the narrowest ship-channel between Long Island and New York. If a wall were built across Hell Gate, the water on one side would sometimes be five feet above that on the other. In the absence of such a wall, the current surges with great rapidity under the Brooklyn Bridge, alternately in opposite directions.

PRACTICAL QUESTIONS.

1. Two faucets, one 8 feet and the other 4 feet below the surface of the water in a cistern, are kept open for a minute. How many times as much water can be drawn from the first as the second?

2. How much water will be discharged per second from a short pipe having a diameter of 4 inches and a depth of 48 feet below the surface of the water?

3. When we pour molasses from a jug, why is the stream so much larger near the nozzle than at some distance from it?

4. Ought a faucet to extend into a barrel beyond the staves?

5. What would be the effect if both openings in one of the arms of Barker's Mill were on the same side?

III. PNEUMATICS.

PNEUMATICS treats of the general properties and the pressure of gases. Since the molecules move among themselves more freely even than those of liquids, the conclusions which we have reached with regard to transmission of pressure, buoyancy, and specific gravity apply also to gases. Since air is the most abundant gas, it is taken as the type of the class, just as water is of liquids.

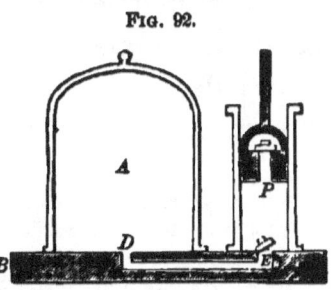

FIG. 92.

The Air-pump.

1. The Air-pump is shown in its essential features in Fig. 92. A is a glass receiver standing on an oiled pump-plate. The tube D connecting the receiver with the cylinder, is closed by the valve E, opening upward. There is a second valve, P, in the piston, also opening upward. Suppose the piston is at the bottom and both valves shut. Let it now be raised, and a vacuum will be produced in the cylinder; the expansive force of the atmosphere in the receiver will open the valve E and drive the air through to fill this empty space. When the piston descends, the valve E will close, while the valve P will open, and the air will pass up above the piston. On elevating the piston a second time, this air is removed from the cylinder, while the air from the receiver passes through as before. At each stroke a portion of the atmosphere is drawn off; but the expansive force becomes less and less, until finally it is insufficient to lift the valves. For this reason a perfect vacuum can not be obtained.

2. The Condenser, in construction, is the same as the air-pump, except that the valve opens inward instead of outward. Instead of exhausting, it forces more air into a vessel.*

3. Properties of Air.—(1.) WEIGHT.—Exhaust the air from a flask which holds 100 cubic inches, and then balance it. On turning the stop-cock, the air will rush in with a whizzing noise and the flask

---

* The practical applications of this pump are numerous. The soda manufacturer uses it to condense carbonic acid in soda-water reservoirs.—The engineer employs it in laying the foundations of bridges. Large tubes or *caissons* are lowered to the bed of the stream, and air being forced in, drives out the water. The workmen are let into the caissons by a sort of trap, and work in this condensed atmosphere.—Pneumatic dispatch-tubes contain a kind of train holding the mail, and back of this a piston fitting the tube. Air is forced in behind the piston or exhausted before it, and so the train is driven through the tube at a high speed.—In the Westinghouse air-brake, condensed air is forced along a tube running underneath the cars, and by its elastic force drives the brakes against the wheel.

will descend (Fig. 93). It will require 31 grains or more to restore the equipoise.

(2.) ELASTICITY is shown in a pop-gun. We compress the atmosphere in the barrel until the elastic force drives out the stopper with a loud report. As we crowd down the piston we feel the elasticity of the air yielding to our strength, like a bent spring.—The bottle-imps, or Cartesian divers, illustrate the same property. Fig. 94 represents a simple form of this apparatus. The cover of a fruit-jar is fitted with a tube, which is inserted in a syringe-bulb.

Fig. 93.

Weighing Air.

Fig. 94.

Cartesian Diver.

The jar is filled with water and the diver placed within. This is a hollow image of glass, having a small opening at the end of the curved tail. If we squeeze the bulb, the air will be forced into the jar and the water will transmit the pressure to the air in the image. This being compressed, more water will enter, and the diver, thus becoming heavier, will descend. On relaxing the grasp of the hand on the bulb, the air will return into it, the air in the image will expand, by its elastic force driving out the water, and the diver, thus lightened of his ballast, will ascend. The nearer the image is to the bottom, the less force will be required to move it. With a little care

132   PRESSURE OF LIQUIDS AND GASES.

it can be made to respond to the slightest pressure, and will rise and fall as if instinct with life.*

Fig. 95.

Expansibility of Air.

(3.) EXPANSIBILITY.—Let a well-dried bladder be partly filled with air and tightly closed. Place it under the receiver and exhaust the air. The air in the bladder expanding will burst it into shreds.

Take two bottles partly filled with colored water. Let a bent tube be inserted tightly in A and loosely in B. Place this apparatus under the receiver and exhaust the air. The expansive force of the air in A will drive the water over into B. On readmitting the air into the receiver, the pressure will return the water into A. It may thus be driven from bottle to bottle at pleasure.†

Fig. 96.

Transfer of Liquid under Receiver.

Hiero's fountain acts on the same principle, as may

---

* This experiment shows also the buoyant force of liquids, their transmission of pressure in every direction, and the increase of the pressure in proportion to the depth. The elasticity of the air, as well as the principles explained by the Cartesian diver, Fig. 94, may be illustrated in the following simple manner: Fill with water a wide-mouth 8-oz. bottle, and also a tiny vial, such as is used by homeopathists. Invert the vial and a few drops of water will run out. Now put it inverted into the bottle, and if it does not sink just below the surface and there float, take it out and add or remove a little water, as may be needed. When this result is reached, cork the bottle so that the cork touches the water. Any pressure on the cork will then be transmitted to the air in the vial, as in the image in Fig 94.

† Prick a hole in the small end of an egg, and place the egg with the big end up in a wine-glass. On exhausting the receiver, the bubble of air in the upper part of the egg will drive the contents down into the glass, and on admitting the air they will be forced back again.

be seen by an examination of Fig. 97. Having removed the jet-tube, the upper globe is partly filled with water. The tube being then replaced, water is poured into the basin on top. The liquid runs down the pipe at the right, into the lower globe. The air in that globe is driven up the tube at the left into the upper globe, and by its elasticity forces the water there out through the jet-tube, forming a tiny fountain.

Fig. 97.

Hiero's Fountain.

Fig. 98.

Hand-glass.

4. Pressure of the Air.—(1.) THE PROOF.— If we cover a hand-glass with one hand, as in Fig. 98, on exhausting the air we shall find the pressure painful.* Tie over one end of the glass a piece of wet bladder. When dry, exhaust the air, and the membrane will burst with a sharp report.†

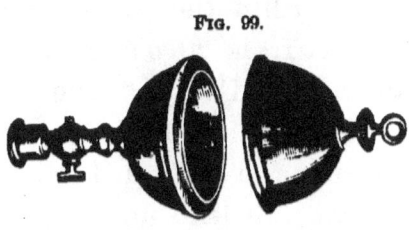

Fig. 99.

Magdeburg Hemispheres.

---

* The exhaustion of the air does not *produce* the pressure on the hand; it simply *reveals* it. The average pressure on each person is 16 tons. It is equal, however, on all parts of the body, and is counteracted by the air within. Hence we never notice it. Persons who go up high mountains or go down in diving-bells feel the change in the pressure.

† To show the crushing force of the atmosphere, take a tin cylinder 15 inches long and 4 inches in diameter. Fit one end with a stop-cock for the exit of the steam. Put in a little water and boil. When the air is entirely

134 PRESSURE OF LIQUIDS AND GASES.

The *Magdeburg Hemispheres* are named from the city in which Guericke, their inventor, resided. They consist of two small brass hemispheres, which fit closely together, but may be separated at pleasure. If, however, the air be exhausted from within, several persons will be required to pull them apart.* In whatever position the hemispheres are held, the pressure is the same.

Fig. 100.

Water held up by Pressure of Air.

(2.) UPWARD PRESSURE.—Fill a tumbler with water, and then lay a sheet of paper over the top. Quickly invert the glass, and the water will be supported by the upward pressure of the air.—Within the glass cylinder, Fig. 101, is a piston working air-tight. Connect the nozzle above with the air-pump by means of a rubber tube and exhaust the air. The weight will leap up as if caught by a spring.

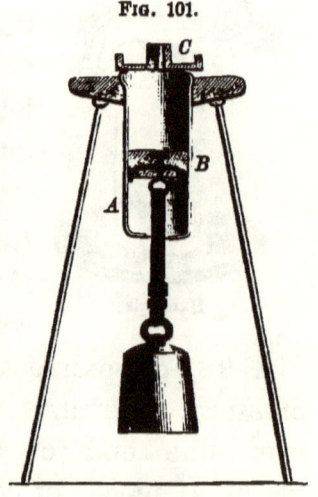

Fig. 101.

The Weight-lifter.

driven out, turn the stop-cock. Pour cold water over the outside to condense the steam, when the cylinder will collapse as if struck by a heavy blow.

* In the museum at Berlin the hemispheres used by Guericke in his experiments are preserved. They are of copper, and 22 inches interior diameter, with a flange an inch wide, making the entire diameter 2 feet. Accompanying is a Latin book by the burgomaster describing numerous pneumatic experiments which he had performed, and containing a woodcut representing **three spans of horses** on each side trying to separate the hemispheres.

(3.) **BUOYANT FORCE OF THE AIR.**—The law of Archimedes (p. 114) holds true in gases. A hollow sphere of glass or copper, Fig. 102, is balanced in the air by a solid lead weight, but on being placed

Fig. 102.

Buoyancy of Air.

Fig. 103.

Torricelli's Experiment.

under the receiver it steadily falls while the air is becoming exhausted. This shows that its weight was partly sustained by the buoyant force of the air.

(4.) AT SEA-LEVEL THE PRESSURE OF THE AIR SUSTAINS A COLUMN OF MERCURY 30 INCHES HIGH, OR OF WATER NEARLY 34 FEET HIGH, AND IS NEARLY 15 LBS. PER SQUARE INCH. Take a strong glass tube about three feet in length, and tie over one end a piece of wet bladder. When dry, fill the tube with mercury, and invert it in a cup of the same liquid. The mercury will sink to a height of about 30 inches. If the area across the tube be one square inch, the metal will weigh about 14.7 lbs. The weight of the column of mercury is equal to the downward pressure on each square inch of the surface of the mercury in the cup. Hence we conclude that the pressure of the atmosphere is 14.7 lbs. per square inch, and will balance a column of mercury 30 inches high. As water is 13¼ times lighter than mercury, the same pressure would balance a column of that liquid 13½ times higher, or 33¼ feet.*

(5.) PRESSURE OF THE AIR VARIES.† Changes of temperature, moisture, etc., continually vary the density of the air, and change the height of the column of liquid it can support. The pressure also increases with the depth. Hence, in a valley the column of mercury stands higher than on a mountain. The pressure of the atmosphere is 29.92 inches only at the level of the sea, and at the temperature of melting ice at latitude 45°. The variation due to latitude is very slight; that due to temperature is greater, and that due to elevation is greatest. Observations on the barometer at any given station

---

* On account of the unwieldly length of the tube required to exhibit the column of water, it is not easy to verify this. It may, however, be prettily illustrated. Pour on the mercury in the cup (Fig. 103) a little water colored with red ink. Then raise the end of the tube above the surface of the metal, but not above that of the water; this will rise in the tube, the mercury passing down in beautifully-beaded globules. The mercurial column is only 30 inches high, while the water will fill the tube. Finish the experiment by puncturing the bladder with a pin, when the water will instantly fall to the cup below.

† We live at the bottom of an aërial ocean whose depth is greater than that of the deepest sea. Its invisible currents surge round us on every side.

are generally "reduced" to what they would be under the standard conditions just mentioned.

(6.) MARIOTTE'S LAW.*
—Fig. 104 represents a long, bent glass tube with the end of the short arm closed. Pour mercury into the long arm until it rises to the point marked zero.† It stands at the same height in both arms, and there is an equilibrium. The air presses on the mercury in the long arm with a force equal to a column of mercury 30 inches high, and the elastic force of the air confined in the short arm is equal to the same amount. Now pour additional mercury into the long arm until it stands at 30 inches above that in the short arm (Fig. 105), and the pressure is doubled. In the short arm, the air is condensed to one half its

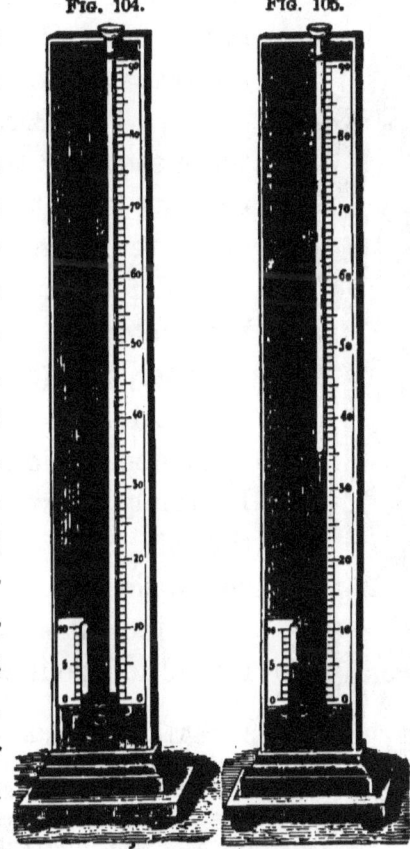

FIG. 104.   FIG. 105.

Mariotte's Tube.

---

* This law was independently discovered by the Englishman, Boyle, and the Frenchman, Mariotte, during the latter part of the seventeenth century. It is often called Boyle's Law.

† By cautiously inclining the apparatus, when a little air will escape, and adding more mercury if needed, the liquid can be made to stand at zero in both arms.

138 PRESSURE OF LIQUIDS AND GASES.

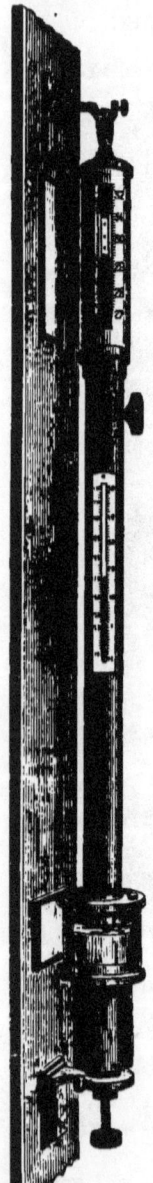

Fig. 106.

The Barometer.

former dimensions, and the elastic force is also doubled.* We therefore conclude that *the elasticity of a gas increases, and the volume diminishes in proportion to the pressure upon it.*

(7.) The BAROMETER is an instrument for measuring the pressure of the air. It consists essentially of the tube and cup of mercury in Fig. 103. A scale is attached for convenience of reference. The barometer is used (*a*) to indicate the weather, and (*b*) to measure the height of mountains.

It does not directly foretell the weather. It simply shows the varying pressure of the air, from which we must draw our conclusions. A continued rise of the mercury indicates fair weather, and a continued fall, foul weather.† Since the press-

---

* The force with which the flying molecules of air beat against the walls of any confining vessel will increase with the diminution of the space through which they can pass. If we give them only half the distance to fly through, they will strike twice as often and exert twice the pressure.

† Mercury is used for filling the barometer because of its weight and low freezing-point. It is said that the first barometer was filled with water. The inventor, Otto von Guericke, erected a tall tube reaching from a cistern in the cellar up through the roof of his house. A wooden image was placed within the tube, floating upon the water. On fine days, this novel weather-prophet would rise above the roof-top and peep out upon the queer old gables of that ancient city, while in foul weather he would retire to the protection of the garret. The accuracy of these movements attracted the attention of the neighbors. Finally, becoming

ure diminishes above the level of the sea, the observer ascertains the fall of the mercury in the barometer, and the temperature by the thermometer; and then, by reference to tables, determines the height.

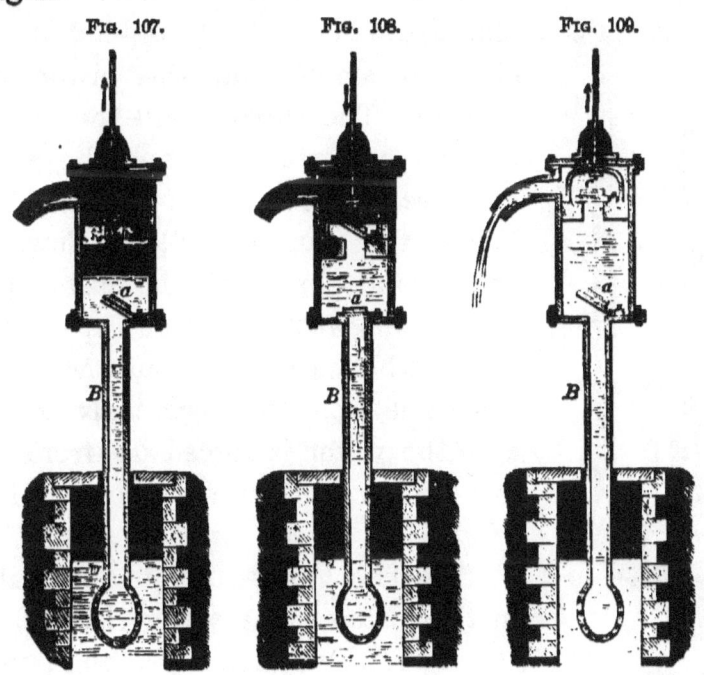

Fig. 107. Fig. 108. Fig. 109.

The Lifting-pump.

5. Pumps.—(1.) The LIFTING-PUMP contains two valves opening upward—one, a, at the top of the *suction-pipe*, B; the other, c, in the piston. Suppose the handle to be raised, the piston being at the bottom of the cylinder and both valves closed. Now

suspicious of Otto's piety, they accused him of being in league with the devil. So the offending philosopher relieved this wicked wooden man from longer dancing attendance upon the weather, and the staid old city was once more at peace.

140 PRESSURE OF LIQUIDS AND GASES.

depress the pump-handle and thereby elevate the piston. This will produce a partial vacuum in the suction-pipe. The pressure of the air on the surface of the water below will force the water up the pipe,

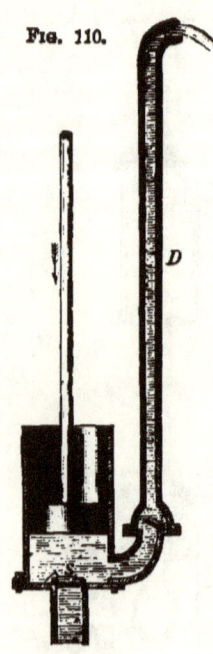

Fig. 110.

Force-pump.

open the valve, and partly fill the chamber. Let the pump-handle be elevated again, and the piston depressed. The valve a will then close, the valve c will open, and the water will rise above the piston (Fig. 108). When the pump-handle is lowered the second time and the piston elevated, the water is lifted up to the spout, whence it flows out; while at the same time the lower valve opens and the water is forced up from below by the pressure of the air (Fig. 109).*

(2.) The FORCE-PUMP has no valve in the piston. The water rises above the lower valve as in the lifting-pump. When the piston descends, the pressure opens the valve in the pipe D, and forces the water up. This pipe may be made of any length, and thus the water driven to any height.

(3.) The FIRE-ENGINE consists of two force-pumps with an air-chamber. The water is driven by the

* If the valves and piston were fitted air-tight, the water could be raised 34 feet (more exactly 13½ times the height of the barometric column) to the lower valve, but owing to various imperfections it commonly reaches about 28 feet. For a similar reason we sometimes find a dozen strokes necessary to "bring water."

PNEUMATICS. 141

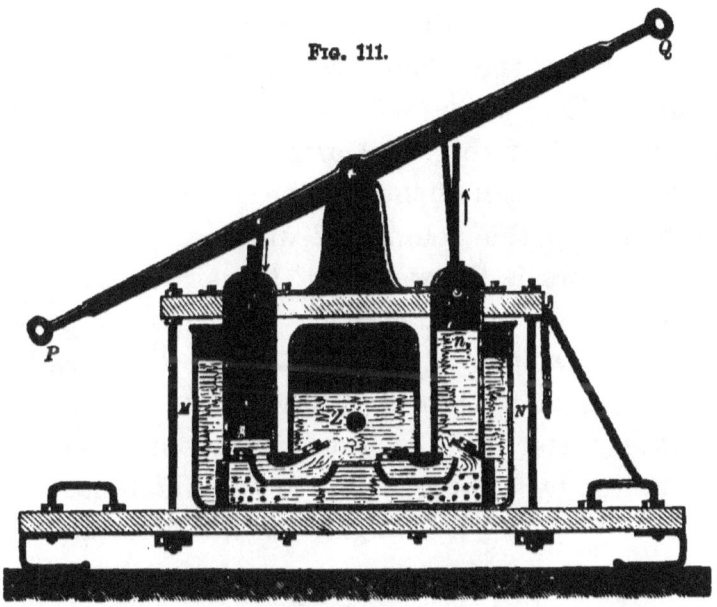

Fig. 111.

Fire-engine.

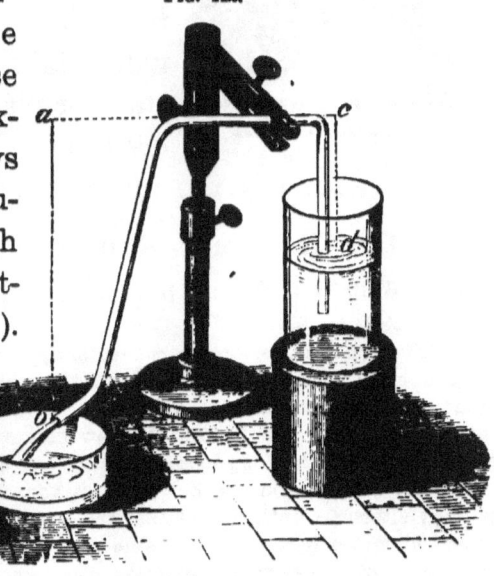

Fig. 112.

Siphon.

pistons *m*, *n*, alternately, into the chamber *R*, whence the air, by its expansive force, throws it out in a continuous stream through the hose-pipe attached at *Z* (Fig. 111).

6. **The Siphon** is a U-shaped tube, having one arm longer than

the other. Insert the short arm in the water, and then applying the mouth to the long arm, exhaust the air. The water will flow from the long arm until the end of the short arm is uncovered.

THEORY OF THE SIPHON.—The pressure of the air at b holds up the column of water $a\ b$, and the upward pressure is the weight of the air less the weight of the column of water $a\ b$. The upward pressure at d is the weight of the air minus the weight of the column of water $c\ d$. Now $c\ d$ is less than $a\ b$, and the water in the tube is driven toward the longer arm by a force equal to the difference in the weight of the two arms.*

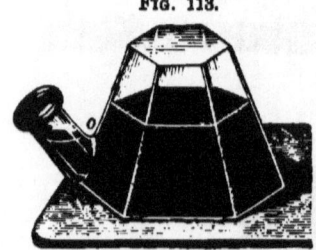

FIG. 113.

Pneumatic Inkstand.

7. The Pneumatic Inkstand can be filled only when tipped so that the nozzle is at the top. The pressure of the air will retain the ink when the stand is placed upright. When used below o, a bubble of air passes in, forcing the ink into the nozzle.

8. The Hydraulic Ram is a machine for raising water where there is a slight fall. The water enters

---

* The siphon is used more conveniently if two tubes of glass or metal are connected with a flexible tube of India rubber. An instructive experiment can then be made if we allow the water to run from one tumbler into another until just before the flow ceases; then quickly elevate the glass containing the long arm, carefully keeping both ends of the siphon under the water, when the flow will set back to the first tumbler. Thus we may alternate until we see that the water flows to the lower level, and ceases whenever it reaches the same level in both glasses. It will add to the beauty of this as well as of many other experiments, to color the water in one tumbler with a few scales of magenta, or with red ink.

through the pipe *A*, fills the reservoir *B*, and lifts the valve *D*. As that closes, the shock raises the

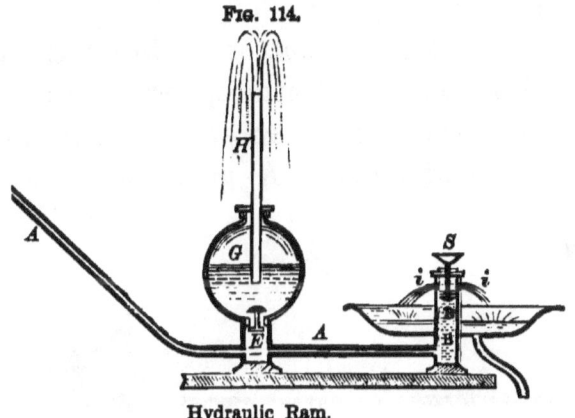

Fig. 114.

Hydraulic Ram.

valve *E* and drives the water into the air-chamber *G*. *D* falls again as soon as an equilibrium is restored. A second shock follows, and more water is thrown into *G*. When the air in *G* is sufficiently condensed, its elastic force drives the water through the pipe *H*.

Fig. 115.

Atomizer.

9. The Atomizer is used to turn a liquid into spray. The blast of air driven from the rubber bulb as it passes over the end of the upright tube, sweeps along the neighboring molecules of air and produces a partial vacuum in the tube.* The

---

* In locomotives, this principle of the adhesion of gases to gases is applied to produce a draft. The waste steam is thrown into the smoke-pipe,

pressure of the air in the bottle drives the liquid up the tube, and at the mouth the blast of air carries it off in fine drops.

The action of a current of air in dragging along with it the adjacent still atmosphere and so tending to produce a vacuum, is shown by the apparatus represented in Fig. 116. A globe, *a*, is connected

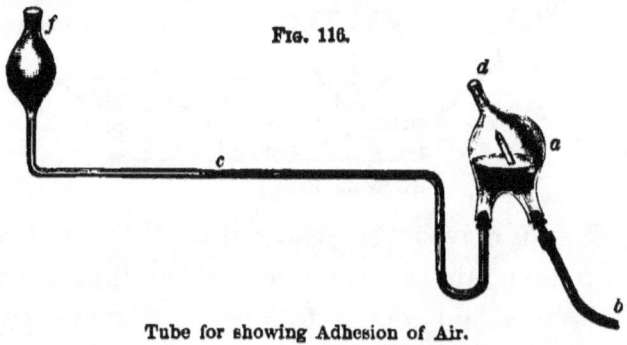

Tube for showing Adhesion of Air.

with a horizontal tube, *c*, containing colored water. Close the opening *d* with the finger, and with the mouth at *b* draw the air out of the globe. A slight rarefaction will cause the liquid, by the pressure of the air at the opening *f*, to be forced into *a*. Now,

and this current sweeps off the smoke from the fire, while the pressure of the atmosphere outside forces the air through the furnace and increases the combustion.—A familiar illustration may be devised by taking two disks of card-board, the lower one fitted with a quill, and the upper one merely kept from sliding off by a pin thrust through it and extending into the quill. The more forcibly air is driven through the quill against the upper disk, the more firmly it will be held to its place. See article "Ball Paradox," in "Popular Science Monthly," April, 1877.—Faraday used to illustrate the principle thus: Hold the hand out flat with the fingers extended and pressed together. Place underneath a piece of paper two inches square. Blow through the opening between the index and the middle finger, and so long as the current is passing the paper will not fall.

if, instead of drawing the air out at b, a jet of air be forced through the tube and out at d, the same effect will be produced.

10. Height of the Atmosphere.—Three opposing forces act upon the air, viz.: gravity, which binds it to the earth, and the centrifugal and repellent forces, which tend to hurl it into space. There must be a point where these balance. At the height of 3.4 miles the mercury in the barometer stands at 15 inches, indicating that half the atmosphere is within about $3\frac{1}{4}$ miles of the earth's surface. Beyond a height of 40 miles the quantity of air is too small to be perceptible in any way.* If it were every-where as dense as it is at sea-level, the upper limit of our atmosphere would be about five miles high.

PRACTICAL QUESTIONS.

[*In these questions, assume the standard conditions mentioned on p. 136.*]

1. Why must we make two openings in a barrel of cider when we tap it?
2. What is the weight of 10 cubic feet of air?
3. What is the pressure of the air on 1 square rod of land?
4. What is the pressure on a pair of Magdeburg hemispheres 4 inches in diameter?
5. How high a column of water can the air sustain when the barometric column stands at 28 inches?
6. If we should add a pressure of two atmospheres (30 lbs. to the square inch), what would be the volume of 100 cubic inches of common air?

---

* In mountain climbing, or ascending to a great height in a balloon, the voyager is apt to suffer on account of the decrease in density of the air. In 1862, Mr. Glaisher ascended nearly 7 miles, and there fainted. His assistant was barely able to open the valve and cause the balloon to descend.

146   PRESSURE OF LIQUIDS AND GASES.

7. If, while the water is running through the siphon, we quickly lift the long arm, what is the effect on the water in the siphon? If we lift the entire siphon?

8. When the mercury stands at 29½ inches in the barometer, how high above the surface of the water can we place the lower pump-valve?

9. Can we raise water to a higher level by means of a siphon?

10. If the air in the chamber of a fire-engine be condensed to $\frac{1}{A}$ its former bulk, what will be the pressure due to the expansive force of the air on every square inch of the air-chamber?

11. What causes the bubbles to rise to the surface when we put a lump of loaf-sugar in hot tea?

12. When will a balloon stop rising? What weight can it lift?

13. The rise and fall of the barometric column shows that the air is lighter in foul and heavier in fair weather. Why is this? *Ans.* Vapor of water is only half as heavy as dry air. When there is a large quantity present in the atmosphere, displacing its own volume of air, the weight of the atmosphere will be correspondingly diminished.

14. When smoke ascends in a straight line from chimneys, is it a proof of the rarity or the density of the air?

15. Explain the action of the common leather-sucker.

16. Did you ever see a bottle really empty?

17. Why is it so tiresome to walk in miry clay? *Ans.* Because the upward pressure of the air is removed from our feet.

18. How does the variation in the pressure of the air affect those who ascend lofty mountains? Who descend in diving-bells?

19. Explain the theory of "sucking cider" through a straw.

20. Would it make any difference in the action of the siphon if the limbs were of unequal diameter?

21. What would be the effect of making a small hole in the top of a diving-bell while in use?

22. The pressure of the atmosphere being 1.03 kg. per sq. cm., what is the amount on 10 sq. meters?

SUMMARY.

Hydrostatics treats of the laws of equilibrium in liquids. Pressure is transmitted by liquids equally in every direction. Water thus becomes a "mechanical power," as in the "Hydraulic Press." Liquids acted on by their weight only, at the same depth, press downward, upward, and sidewise with equal force. This pressure is independent of the size of the vessel, but increases with the depth. Wells, springs, aqueducts, fountains, and the water-supply of cities illustrate the tendency of

SUMMARY. 147

water to seek its level. The ancients understood this law, but had no suitable material for making the immense pipes needed; just so the art of printing awaited the invention of paper. Specific gravity, or the relative weights of the same volume of different substances, is found by comparing them with the weight of the same volume of water. This is easily done, since, according to the law of Archimedes, a body immersed in water is buoyed up by a force equal to the weight of the water displaced; *i. e.*, it loses in weight an amount equal to that of the same volume of water. Hence spec. grav. $= \dfrac{\text{weight in vacuum}}{\text{weight in vacuum} - \text{weight in water}}$. A floating body displaces only its own weight of liquid. This explains the buoyancy which supports a ship, why a floating log is partly out of water, and many similar phenomena.

Hydrodynamics treats of moving liquids. The laws of falling bodies in theory apply; so that a descending jet of water will acquire the same velocity that a stone would in falling to the ground from the surface of the water; and an ascending jet would need to have the same velocity in order to reach that height. The quantity of water discharged through any orifice equals the area of the opening multiplied by the velocity of the stream. The chief resistance to the motion of a liquid is the friction of the air and against the sides of the pipe, and, in the case of rivers, against the banks and bottom of the channel. The force of falling water is utilized in the arts by means of water-wheels. There are four kinds—overshot, undershot, breast, and turbine. The principles of wave-motion, so essential to the understanding of sound, light, etc., are most easily studied in connection with water. A stone let fall into a quiet pool sets in motion a series of concentric waves, whose particles move in ellipses, while the movement passes to the outermost edge of the water, and is then transmitted to the ground beyond. The velocity of the particles is much less than that of the wave itself. A handful of stones acts in the same way, but sets in motion many series of waves. Hence arise the phenomena of interference.

Pneumatics treats of the properties and the laws of equilibrium of gases. The air being composed of matter, has all the properties we associate with matter, as weight, indestructibility,

148   PRESSURE OF LIQUIDS AND GASES.

extension, compressibility, etc. The elasticity of the air, according to Mariotte's law, is inversely proportional to its volume, and this is inversely proportional to the pressure upon the air; both heat and pressure increasing the elasticity of a gas. The air, like other fluids, transmits the weight of its own particles, as well as any outside pressure, equally in every direction; hence the upward pressure or buoyant force of the atmosphere. A balloon rises because it is buoyed up by a force equal to the weight of the air it displaces. It floats in the air for the same reason that a ship floats on the ocean. When smoke falls it is heavier than the surrounding atmosphere. When it rises, it is carried up by adhesion of warm air, which is lighter than that surrounding the current. The air-pump is used for exhausting the air from, and the condenser for condensing the air into, a receiver. A vacuum in which there remains only $\frac{1}{1000000}$ of the atmosphere can be obtained by means of Sprengel's air-pump, which acts on the principle of the adhesion of the air to a column of falling mercury. The average pressure of the air being 15 lbs. to the square inch, equals that of a column of water 34 feet, and of mercury 30 inches or 760 millimeters high. This amount varies incessantly through atmospheric changes caused by alterations in the wind, heat of the sun, etc. The barometer measures the pressure of the atmosphere, and is used to determine the height of mountains and the changes of the weather. The action of the siphon, the pneumatic inkstand, and of the different kinds of pumps, is based upon the pressure of the air.

HISTORICAL SKETCH.

Hydrostatics is comparatively a modern science. The Romans had a knowledge of the fact that "liquids rise to the level of their source," but they had no means of making iron pipes strong enough to resist the pressure.* They were there-

* The ancient engineers sometimes availed themselves of this principle. Not far from Rachel's Tomb, Jerusalem, are the remains of a conduit once used for supplying the city with water. The valley was crossed by means of an inverted siphon. The pipe was about two miles long and fifteen

HISTORICAL SKETCH. 149

fore forced to carry water into the Imperial City by means of enormous aqueducts, one of which was 63 miles long, and was supported by arches 100 feet high. The ancient Egyptians and Chaldeans were probably the first to investigate the most obvious laws of liquids from the necessity of irrigating their land. Archimedes, in the third century B.C., invented a kind of pump called *Archimedes' Screw*, demonstrated the principle of equilibrium, known now as "*Archimedes' Law*" (p. 114), and found out the method of obtaining the specific gravity of bodies. The discovery of the last is historical. Hiero of Syracuse suspected that a gold crown had been fraudulently alloyed with silver. He accordingly asked Archimedes to find out the fact without injuring the workmanship of the crown. One day going into a bath-tub full of water, the thought struck the philosopher that as much water must run over the side as was equal to the volume of his body. Electrified by the idea, he sprang out and ran through the streets, shouting: "Eureka!" (I have found it!)

The ancients never dreamed of associating the air with gross matter. To them it was the spirit, the life, the breath. Noticing how the atmosphere rushes in to fill any vacant space, the followers of Aristotle explained it by saying, "Nature abhors a vacuum." This principle answered the purpose of philosophers for 2,000 years. In 1640, some workmen were employed by the Duke of Tuscany to dig a deep well near Florence. They found to their surprise that the water would not rise in the pump as high as the lower valve. More disgusted with nature than nature was with the vacuum in their pump, they applied to Galileo. The aged philosopher answered—half in jest, we hope, certainly he was half in earnest—"Nature does not abhor a vacuum beyond 34 feet." His pupil, Torricelli, however, discovered the secret. He reasoned that there is a force which holds up the water, and as mercury is $13\frac{1}{2}$ times as heavy as water, it would sustain a column of that liquid only 34 feet $\div 13\frac{1}{2} = 30$ inches high. Trying the experiment shown in Fig. 103, he verified the conclusion that the weight of the air is the unknown force. But the opinion was not generally received.

inches in diameter. It consisted of perforated blocks of stone, ground smooth at the joints, and fastened with a hard cement.

150 PRESSURE OF LIQUIDS AND GASES.

Pascal next reasoned that if the weight of the air is really the force, then at the summit of a high mountain it is weakened, and the column would be lower. He accordingly carried his apparatus to the top of a tower, and finding a slight fall in the mercury, he asked his brother-in-law, Perrier, who lived near Puy de Dôme, a mountain in Southern France, to test the conclusion. On trial, it was found that the mercury fell 3 inches. "A result," wrote Perrier, "which ravished us with admiration and astonishment." Thus was discovered the germ of our modern barometer, and the dogma of the philosophers soon gave place to the law of gravitation and our present views concerning the atmosphere.

Consult Pepper's "Cyclopedic Science"; Bert's "Atmospheric Pressure and Life," in "Popular Science Monthly," Vol. XI., p. 316; "Appleton's Cyclopedia," Articles on Hydromechanics, Atmosphere, Pneumatics, etc.; Delaunay, "Mécanique Rationelle"; Boutan et D'Almeida, "Cours de Physique"; Müller, "Lehrbuch der Physik und Meteorologie."

On the theory of Wave-motion, and the subjects of Sound and Light, which are now to follow, consult Lockyer's "Studies in Spectrum Analysis"; Lloyd's "Wave Theory"; Taylor's "Science of Music"; Blaserna's "Theory of Sound in Relation to Music"; Tyndall's "Sound" and "Light"; Lockyer's "Water-waves and Sound-waves" in "Popular Science Monthly," Vol. XIII., p. 166; Shaw's "How Sound and Words are Produced," in "Popular Science Monthly," Vol. XIII., p. 43; Mayer on "Sound"; Schellen's "Spectrum Analysis"; Airy's "Optics"; Lockyer's "Spectroscope"; Chevreul's "Colors"; Spottiswoode's "Polarization of Light"; Lommel's "Nature of Light"; Helmholtz's "Popular Lectures on Scientific Subjects"; "Appleton's Cyclopedia," Articles on Sound, Light, Spectrum, Spectrum Analysis, Spectacles, Heat, etc.; Stokes' "Absorption and Colors," and Forbes' "Radiation," in "Science Lectures at South Kensington," Vol. I.; Mayer and Barnard's "Light"; Draper's "Popular Exposition of some Scientific Experiments," in "Harper's Magazine" for 1877; Core's "Modern Discoveries in Sound," in Manchester Science Lectures, '77-8; Dolbear's "Art of Projecting"; Draper's "Scientific Memoirs"; Steele's "Physiology," Section on Sight, pp. 187-196.

VI.

ON SOUND.

SCIENCE ought to teach us to see the invisible as well as the visible in nature: to picture to our mind's eye those operations that entirely elude the eye of the body; to look at the very atoms of matter, in motion and in rest, and to follow them forth into the world of the senses."—TYNDALL.

ANALYSIS OF SOUND.

ACOUSTICS, OR THE SCIENCE OF SOUND.

1. **Production of Sound.**
2. **Transmission of Sound.**
   - (1.) Through Air.
   - (2.) In a Vacuum.
   - (3.) In Liquids.
   - (4.) In Solids.
   - (5.) Production of Motion by Sound.
   - (6.) Co-vibration through Air as a Medium.
   - (7.) Velocity of Transmission.
   - (8.) Loudness of Sound.
3. **Refraction of Sound.**
4. **Reflection of Sound.**
   - (1.) Law of Reflection.
   - (2.) Echoes.
   - (3.) Decrease by Reflection.
   - (4.) Acoustic Clouds.
5. **Musical Sounds.**
   - (1.) Difference between Noise and Music.
   - (2.) Pitch.
   - (3.) The Siren.
   - (4.) Wave-lengths.
   - (5.) Tones in Unison.
6. **Interference of Sound.**
7. **Vibration of Cords.**
   - (1.) The Sonometer.
   - (2.) Laws of Vibration.
   - (3.) Nodes.
   - (4.) Acoustic Figures.
   - (5.) Harmonics.
   - (6.) Nodes of a Bell.
   - (7.) Nodes of a Sounding-board.
   - (8.) Musical Scale.
8. **Vibration of Columns of Air.**
9. **Wind Instruments.**
10. **Co-vibration.**
    - (1.) Sensitive Flames.
    - (2.) Singing Flames.
11. **The Phonograph.**
12. **The Ear.**
    - (1.) Range of the Ear.
    - (2.) Ability to Analyze Sound.

Acoustics, or the Science of Sound.*

1. Production of Sound.—By lightly tapping a glass fruit-dish, we can throw the sides into motion visible to the eye.—Fill a goblet half-full of water, and rub a wet finger lightly around the upper edge of the glass. The sides will vibrate, and cause tiny waves to ripple the surface of the water.—Hold a card close to the prongs of a vibrating tuning-fork, and you can hear the repeated taps. Place the cheek near them, and you will feel the little puffs of wind. Insert the handle between your teeth, and you will experience the indescribable thrill of the swinging metal. The

FIG. 117.

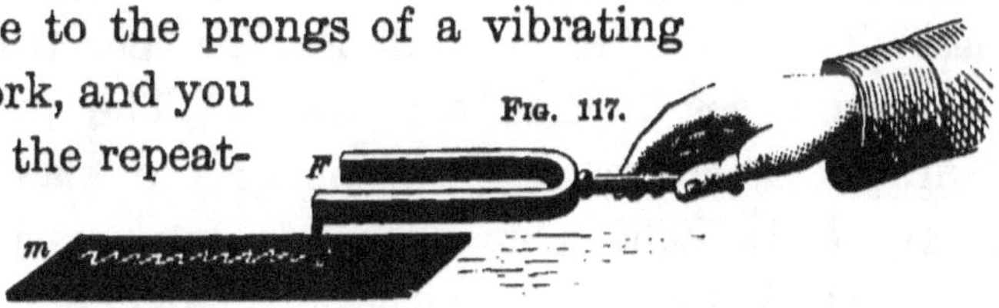

Tuning-fork Registering its Vibrations.

---

* The term *sound* is used in two senses—the *subjective* (which has reference to our mind) and the *objective* (which refers to the objects around us). (1.) Sound is the sensation produced upon the organ of hearing by vibrations in matter. In this use of the word there can be no sound where there is no ear to catch the vibrations.—An oak falls in the forest, and if there is no ear to hear it there is no noise, and the old tree drops quietly to its resting-place.—Niagara's flood poured over its rocky precipice for ages, but fell silently to the ground. There were the vibrations of earth and air, but there was no ear to receive them and translate them into sound. When the first foot trod the primeval solitude, and the ear felt the pulsations from the torrent, then the roaring cataract found a voice and broke its lasting silence. (2.) Sound consists of those vibrations of matter

tuning-fork may be made to draw the outline of its vibrations upon a smoked glass. Fasten upon one prong a sharp point, and drawing the fork along, a sinuous line will show the width (amplitude) of the vibrations.

2. Transmission of Sound. (1.) THROUGH AIR. In order that any medium shall transmit sound, it must be elastic. Most known bodies possess some elasticity, and hence sound may be transmitted through gases, liquids, and solids. The prong of a tuning-fork advances, condensing the elastic air in front of it. This transmits the compression to the air next forward, while the fork swings backward, leaving a rarefaction next to the compression. This

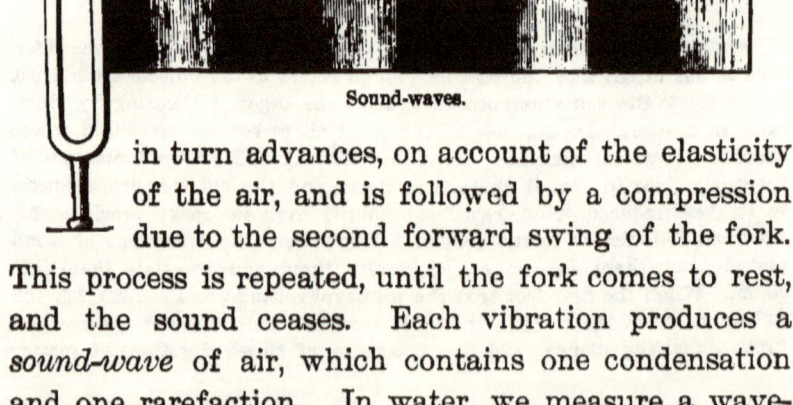

Sound-waves.

in turn advances, on account of the elasticity of the air, and is followed by a compression due to the second forward swing of the fork. This process is repeated, until the fork comes to rest, and the sound ceases. Each vibration produces a *sound-wave* of air, which contains one condensation and one rarefaction. In water, we measure a wave-

---

capable of producing a sensation upon the organ of hearing. In this use of the word there can be a sound in the absence of the ear. An object falls and the vibrations are produced, though there may be no organ of hearing to receive an impression from them. This is the sense in which the term sound is used in Physics.

TRANSMISSION OF SOUND. 155

length from crest to crest; in air, from condensation to condensation. The condensation of the sound-wave corresponds to the crest, and the rarefaction of the sound-wave to the hollow of the water-wave. In Fig. 118, the dark spaces a, b, c, d represent the condensations, and a', b', c' the rarefactions; the wave-lengths are the distances ab, bc, cd.

If we fire a gun, the gases which are produced expand suddenly and force the air outward in every

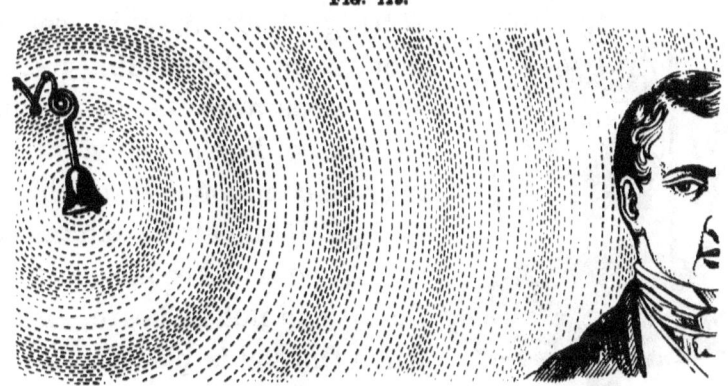

Fig. 119.

Propagation of Sound.

direction. This hollow shell of condensed air imparts its motion to the next one, while it springs back by its elasticity and becomes rarefied. The second shell rushes forward with the motion received, then bounds back and becomes rarefied. Thus each shell of air takes up the motion and imparts it to the next. The wave, consisting of a condensation and a rarefaction, proceeds onward. It is, however, as in water-waves, a movement of the *form* only, while the particles vibrate but a short distance to and fro.

The molecules in water-waves oscillate *vertically;* those in sound-waves *horizontally*, or parallel to the line of motion.*

Fig. 120.

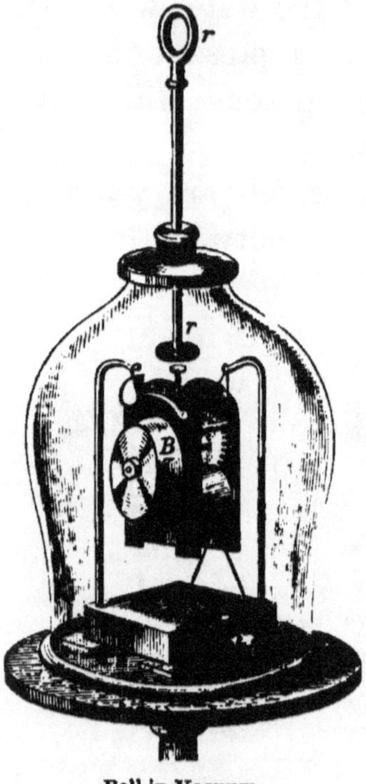

If a bell be rung, the adjacent air is set in motion; thence, by a series of condensations and rarefactions, the vibrations are conveyed to the ear.†

(2.) IN A VACUUM. The bell *B* (Fig. 120) may be set in motion by the sliding-rod *r*. The apparatus is suspended by silk cords, that no vibration may be conducted through the pump. If the air be exhausted, the sound will become so faint that it can not be heard, except when the ear is placed close to the receiver.‡

Bell in Vacuum.

In very elevated regions sounds are diminished in loudness, and it is difficult

---

* A continuous blast of air produces no sound. The rush of the grand aërial rivers above us we never hear. They flow on in the upper regions ceaselessly but silently. Let, however, the great billows strike a tree and wrench it from the ground, and we can hear the secondary, shorter waves which set out from the struggling limbs and the tossing leaves.

† "It is marvelous," says Youmans, "how slight an impulse throws a vast amount of air into motion. We can easily hear the song of a bird 500 feet above us. For its melody to reach us it must have filled with wave-pulsations a sphere of air 1,000 feet in diameter, or set in motion eighteen tons of the atmosphere."

‡ There would be perfect silence in a perfect vacuum. No sound is transmitted to the earth from the regions of space. The movements of the heavenly bodies are noiseless.

to carry on a conversation. The reverse takes place in deep mines and diving-bells.

(3.) IN LIQUIDS. Let two persons immerse themselves in water at a distance of twenty or thirty yards from one another. If one of them strikes two pebbles together, or rings a bell, the other will hear the sound with the utmost clearness.

(4.) IN SOLIDS. The "Lovers' Telephone" consists of a pair of little cups, the bottom of each being made of an elastic substance, like stretched bladder, and connected with that of the other by a string. By stretching the string elastic force is developed, and on talking into one of the cups the sound is readily heard at the other. On relaxing the string and thus diminishing the elasticity, sound ceases to be conveyed perceptibly by it. By putting the ear against the ground, one may hear the tread of footsteps that are inaudible through the air alone.*

(5.) PRODUCTION OF MOTION BY SOUND. If a tuning-fork be excited and its stem be pressed firmly against a table, the sound will become much louder. The solid fork communicates its vibration to the table, which in turn gives its vibration to a much larger body of air than that in contact with the fork alone. Tuning-forks are generally mounted upon resonance boxes, the whole body of air within, as well as the box itself, thus co-vibrating with the fork.

---

* Wheatstone invented a beautiful experiment to show the transmission of sound through wood. Upon the top of a music-box, he rested the end of a wooden rod reaching to the room above, and insulated from the ceiling by India rubber. A violin being placed on the top of the rod, the sounds from the box below filled the upper room, appearing to emanate from the violin.

158　　　ACOUSTICS.

(6.) CO-VIBRATION THROUGH AIR AS A MEDIUM. The air between any source of sound and the ear is like an elastic spring between a pair of tuning-forks of the same size and material. When the prong of the first fork swings from a to b (Fig. 121), a condensation is propagated through the spring and makes

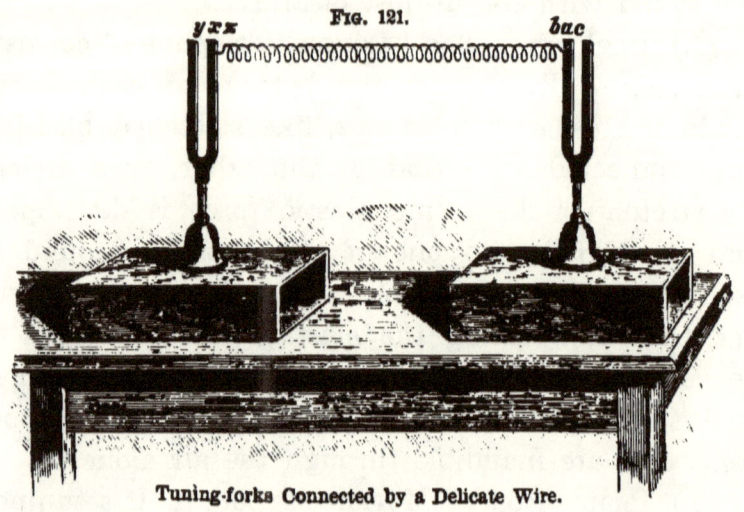

Tuning-forks Connected by a Delicate Wire.

the prong of the second fork swing slightly from x toward y. A rarefaction follows, making it swing from x toward z. The succession of these properly-timed impulses causes an accumulation of energy to be imparted through the air to the second fork, which soon gives forth an audible sound. The elastic bodies composing the ear in like manner accept vibrations from outside, and their motion is perceived as sound.

(7.) THE VELOCITY OF SOUND depends on the ratio of the *elasticity* to the *density* of the medium through which it passes. The higher the elasticity,

TRANSMISSION OF SOUND. 159

the more promptly and rapidly the motion is transmittted, since the elastic force acts like a bent spring between the molecules; and the greater the density, the more molecules to be set in motion, and hence the slower the transmission.

Sound travels through air (at 32° F.) 1,090 feet per second. A rise in temperature diminishes the density of the air, and thus increases the velocity of sound. A difference of 1° F. makes a variation of a little more than one foot. Sound also moves faster in damp than in dry air.

Sound travels through water about 4,700 feet per second. Water being denser than air should on this account conduct sound more slowly; but its high elasticity (p. 10), measured by the amount of force required to compress it, more than quadruples the rate.

Sound travels through solids faster than through air. This may be illustrated by placing the ear close to the horizontal bar at one end of an iron fence, while a person strikes a sharp blow at the other end. Two sounds will reach the ear—one through the metal, and afterward another through the air. The velocity varies with the nature of the solid. In the metals it is from four to sixteen times that in air.

*Different sounds travel with sensibly the same velocity.** A band may be playing at a distance, yet

---

* It has been said that the "heaviest thunder travels no faster than the softest whisper." Mallet, however, found that in blasting with a charge of 2,000 lbs., the velocity was 967 feet per second, while with 12,000 lbs. it was increased to 1,210 feet. Parry in his Arctic travels states that, on a certain occasion, the sound of the sunset-gun reached his ears before the officer's word of command to fire, proving that the report of the cannon traveled sensibly faster than the sound of the voice.

the harmony of the different instruments is preserved. The soft and the loud, the high and the low notes reach the ear at the same time.

Velocity of sound used to find distance. Light travels instantaneously so far as all distances on the earth are concerned. Sound moves more slowly. We see a chopper strike with his ax, and a moment elapses before we hear the blow. If one second intervenes the distance is about 1,090 feet. By means of the second-hand of a watch or the beating of our pulse, we can count the seconds that elapse between a flash of lightning and the peal of thunder which follows. Multiplying the velocity of sound by the number of seconds, we obtain the distance of the thunder-bolt.

(8.) THE LOUDNESS OF SOUND depends chiefly on the amplitude of the vibration, if the air be quiet. The energy of the vibration is proportional to the square of the amplitude, *i. e.*, the arc through which the molecule swings to either side of its position of rest. But loudness is a sensation, and no accurate measurement of sensations has yet been made. The loudness of sound depends also on the density of the air. On the top of a mountain, because of the rare atmosphere, there are fewer molecules to be set in motion, hence the effect on the ear is less intense.

Mechanically considered, the intensity of sound*

---

* The same proportion obtains in Gravitation, Sound, Light, and Heat. We have seen how the motion of the common Pendulum is due to the force of Gravity, and reveals the Laws of Falling Bodies. Now we find that the Pendulum, and even the principles of Reflected Motion and Momentum, are linked with the phenomena of Sound. As we progress further, we shall

diminishes as the square of the distance increases.
The sound-wave expands in the form of a sphere. The larger the sphere, the greater the number of air particles to be set in motion, and the feebler their vibration. The surfaces of spheres are proportional to the squares of their radii; the radii of sound-spheres are their distances from the center of disturbance. Hence the force with which the molecules will strike the ear decreases as the square of our distance from the sounding body increases.

Speaking-tubes conduct sound to distant rooms because they prevent the waves from expanding and losing their intensity.* The *ear-trumpet* collects waves of sound and reflects them into the ear. The *speaking-trumpet* is based on the same principle as the speaking-tube. The sound of the voice is strengthened also by the co-vibration of the walls of the trumpet.

3. Refraction of Sound.—When a sound-wave goes obliquely from one medium to another, it is bent out of its course. Like light, it may be passed through a lens and brought to a focus. In Fig. 122 is shown a bag of thin India rubber or collodion, filled with carbonic acid gas so as to assume the form of a lens. A watch is placed at one focus of this and the ear at the other. The ticks of the watch can be heard, while outside the focus they are inaudible.

find how Nature is thus interwoven every-where with proofs of a common plan and a common Author.

* Biot held a conversation through a Paris water-pipe 3,120 feet long. He says that "it was so easy to be heard, that the only way not to be heard was not to speak at all."

162 ACOUSTICS.

4. Reflection of Sound.—When a sound-wave strikes against the surface of another medium, a portion goes on while the rest is reflected.

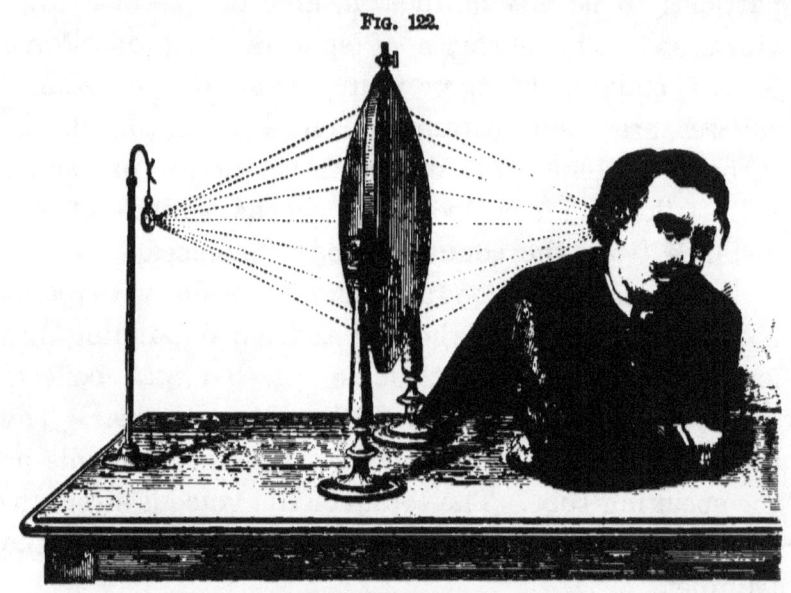

Fig. 122.

Refraction of Sound.

(1.) THE LAW is that of Motion;—the angle of incidence is equal to that of reflection.* If the reflecting surface be very near, the reflected sound will join the direct one and strengthen it. This accounts for the well-known fact that a speaker can be heard

---

* Domes and curved walls reflect sound as mirrors do light. Thus, in the gallery under the dome of St. Paul's Cathedral, London, persons standing close to the wall can whisper to each other and be heard at a great distance.—Two persons, placed with their backs to each other, at the foci of an oval room, or "Whispering Gallery," can carry on a conversation that will be inaudible to spectators standing between them.—The covered recesses on the opposite sides of a street, or the arches of a stone bridge, oftentimes reflect sound so as to enable persons seated at the foci to converse in whispers while loud noises are being made in the open space between these semi-domes.

REFLECTION OF SOUND. 163

more easily in a room than in the open air, and that a smooth wall back of the stand re-enforces the voice. The old-fashioned "sounding-boards" were by no

Fig. 123.

Reflection of Sound.

means inefficient, however singular may have been their appearance.

By revolving a disk of card-board from which a pair of sectors have been cut out, and blowing against it with a trumpet or whistle, a person stationed at the proper angle will notice a beating sound due to successive reflection and transmission of the waves.

(2.) ECHOES are produced where the reflecting surface is so distant that we can distinguish the reflected from the direct sound. If the sound be short and quick, this requires at least fifty or sixty feet; but if it be an articulate one, as in ordinary speech, more than a hundred feet are necessary. It is possible to pronounce and hear distinctly about five syllables in a second; 1,120 ft. (the velocity at a medium temperature) $\div 5 = 224$ ft.* If the wave travel 224 feet in going and returning, the advancing and returning sounds will not blend, and

---

* When several parallel surfaces are properly situated, the echo may be repeated backward and forward in a surprising manner. In Princeton, Ind., there is an echo between two buildings that will return the word "Knickerbocker" twenty times. So many persons visited the place that the city council forbade the use of the echo after 9 o'clock at night.—At Woodstock, England, an echo returns seventeen syllables by day and twenty by night. The reflecting surface is distant about 2,300 feet, and a sharp ha! will come back a ringing ha, ha, ha!—The echo is often softened, as in the Alpine regions, where it warbles a beautiful accompaniment to the shepherd's horn.—The celebrated echo of the Metelli at Rome is said to have been capable of distinctly repeating the first line of the Æneid eight times.—In Fairfax County, Va., is an echo which will return twenty notes played on a flute.—The tick of a watch may be heard from one end of the Church of St. Albans to the other.—At Carisbrook Castle, Isle of Wight, is a well 210 feet deep and twelve feet wide, lined with smooth masonry. When a pin is dropped into the well it is distinctly heard to strike the water.—In certain parts of the Colosseum at London the tearing of paper sounds like the patter of hail, while a single exclamation comes back a peal of laughter.—The dome of the Baptistery of the Cathedral at Pisa has a wonderful echo. During some experiments there, the author found every noise, even the rattle of benches on the pavement below, to be reflected back as if from an immense distance and to return mellowed and softened into music.—An interesting illustration of the reflection of sound is found at the so-called Echo River, of the Mammoth Cave, Ky. Sounding in succession the notes G, E, C, at the middle of the tunnel, the boatman receives the echoes, all mingled into a full chord, for eight or ten seconds afterward.

the ear will be able to detect an interval between them. A person speaking in a loud voice squarely in front of a large smooth wall 112 feet distant, can distinguish the echo of the last syllable he utters; at 224 feet, the last two syllables, etc.

(3.) DECREASE OF SOUND BY REFLECTION.—If we strike the bell, represented in Fig. 120, before a vacuum is produced, we shall find a marked difference between its sound under the glass receiver and in the open air. Floors are deadened with tan-bark or mortar, since as the sound-wave passes from particle to particle of the unhomogeneous mass, it becomes weakened by partial reflection. The air at night is more homogeneous, and hence sounds are heard farther and more clearly than in the day-time.

(4.) ACOUSTIC CLOUDS are masses of moist air of varying density, which act upon sounds as common clouds do upon light, wasting it by repeated reflections. They may exist in the clearest weather. To their presence is to be attributed the variation often noticed in the distance at which well-known sounds, as the ringing of church bells, blowing of engine-whistles, etc., are heard at different times.*

5. **Musical Sounds.**—(1.) THE DIFFERENCE BE-TWEEN NOISE AND MUSIC is that between irregular and

---

* The extinction of sound by such agencies is often almost incredible. Thus two observers looking across the valley of the Chickahominy at the battle of Gaines' Mill failed to hear a sound of the conflict, though they could clearly see the lines of soldiers, the batteries, and the flash of the guns.—These phenomena are ascribed by many to an elevation or a depression of the wave-front so that the sound passes above the observer, or is stopped before it reaches him. See "Stewart's Physics," p. 141.

regular vibrations. Whatever the cause which sets the air in motion, if the vibrations are uniform and rapid enough, the sound is musical. If the ticks of a watch could be made with sufficient rapidity, they would lose their individuality and blend into a musical tone. If the puffs of a locomotive could reach fifty or sixty a second, its approach would be heralded by a tremendous organ-peal.*

(2.) PITCH depends on the rapidity of the vibrations. Thus, if we hold a card against the cogs of a rapidly-revolving wheel, we shall obtain a clear tone; and the faster the wheel turns, the shriller the tone, *i. e.*, the higher the pitch.

(3.) THE NUMBER OF VIBRATIONS PER SECOND is determined by an instrument called the *siren*. It consists of a cylindrical box (Figs. 124 and 125), the top of which is pierced with a series of holes. Over this is a plate with a corresponding series, fixed to a vertical rod, which is pivoted on the lower plate so as to revolve easily. It is provided with an endless screw (Fig. 125), which operates some clock-work.

* The pavement of London is largely composed of granite blocks, four inches in width. A cab-wheel jolting over this at the rate of eight miles per hour produces a succession of 35 sounds per second. These link themselves into a soft, deep musical tone, that will bear comparison with notes derived from more sentimental sources, even though it may seem confused to a hearer in its midst. This tendency of Nature to music is something wonderful. "Even friction," says Tyndall, "is rhythmic." A bullet flying through the air sings softly as a bird. The limbs and leaves of trees murmur as they sway in the breeze. Falling water, singing birds, sighing winds, every-where attest that the same Divine love of the beautiful which causes the rivers to wind through the landscape, the trees to bend in a graceful curve—the line of beauty—and the rarest flowers to bud and blossom where no eye save His may see them, delights also in the anthem of praise which Nature sings for His ear alone.

MUSICAL SOUNDS. 167

On the dial (Fig. 124), we can see the number of turns made by the upper disk. The holes in the two disks are oppositely inclined, so that when a current of air is forced in from below it passes up through the openings in the lower disk, and striking against

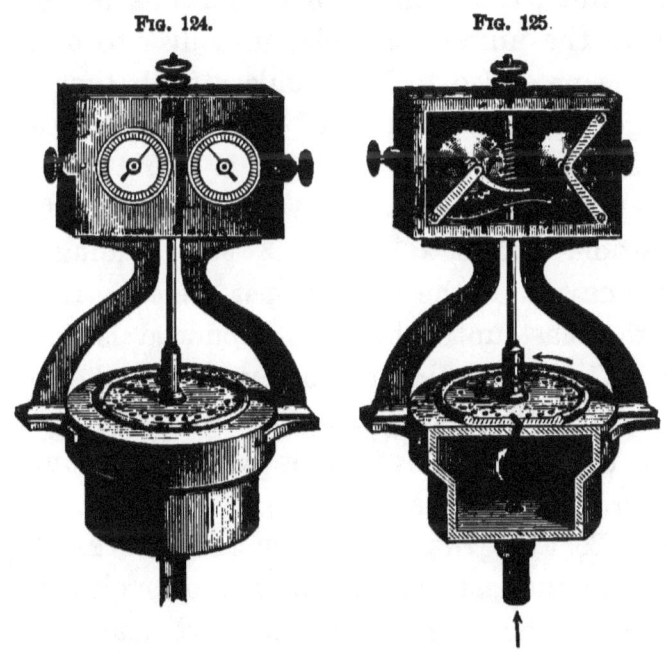

Fig. 124.   Fig. 125.

The Siren.

the sides of those in the upper disk, causes it to revolve. As that turns, it alternately opens and closes the orifices in the lower disk, and thus converts the steady stream of air into uniform puffs. At first they succeed each other so slowly that they may be counted. But, as the motion increases, they link themselves together, and pass into a full, melodious note. As the velocity augments, the pitch rises, until the music becomes painfully shrill. Diminish the speed, and the pitch falls.

To find, therefore, the number of vibrations in a given sound, force the air through the siren until the required pitch is reached. See on the dial, at the end of a minute, the number of revolutions of the disk. Suppose the number of holes in a disk to be 10, and the tone produced to be in unison with that of a C_3 tuning-fork. The number of revolutions indicated on the dial at the end of a minute is found to be 1,536. There were 10 puffs, or 10 waves of sound, for each revolution. $1,536 \times 10 = 15,360$. Dividing this by 60, we have 256, the number per second. Increasing now the blast until the tone produced is in unison with a C_4 tuning-fork, the octave above the first, the number of vibrations per second is found to be 512. Hence the *octave* of a tone is caused by double the number of vibrations.

(4.) To Find the Length of the Wave.—Suppose the air in the last experiment was of such a temperature that the foremost sound-wave traveled 1,120 feet in a second. In that space there were 256 sound-waves. Dividing 1,120 by 256, we have $4\frac{3}{8}$ ft. as the length of each. We thus find the wave-length by dividing the velocity by the number of vibrations per second. As the pitch is elevated by rapidity of vibration, we perceive that the low tones in music are produced by the long waves and the high tones by the short ones.*

---

* The aerial waves are seemingly shortened when the source of sound is approaching, whether by its own motion or the hearer's, and lengthened

(5.) TONES IN UNISON.—If the string of a violin, the cord of a guitar, the parchment of a drum, and the pipe of an organ, produce the same tone, it is because they are executing the same number of vibrations per second. If a voice and a piano perform the same music, the steel strings of the piano and the vocal cords of the singer vibrate together and send out sound-waves of the same length.

6. Interference of Sound-waves. — Just as two water-waves by meeting in opposite phases may destroy one another, so by a proper adjustment two sound-waves may be made to interfere, and, if exactly equal and opposite, to produce silence. Fig. 126 represents a piece of apparatus intended to show this. Let a tone, such as C_3, be sounded in the mouth-piece at a. The waves divide at the end of the first India-rubber tube and reunite on entering the second, before entering the ear at b. One branch of the channel is made of two tubes, one of which slides over the other so that the branch may be lengthened at will. If it be pulled up so high that the waves passing through it shall traverse a half wave-length more in distance than those in the fixed branch, opposite phases will meet where they reunite, and the list-

---

when it is receding. In the former case, the tone of the sound is more acute; in the latter, graver. This is strikingly illustrated when a swift train rushes past a station, the whistle blowing. While the cars are approaching, a person hears a note somewhat sharper; after it has passed, one somewhat flatter than the true note. Still more obvious is the change when two trains pass each other. A person unfamiliar with the arrangement would suppose a different bell was rung. In one case more and in the other fewer waves reach the ears in a second.

ener notices great weakening of the sound. By proper handling, the sound received may be made to become alternately strong and weak without any change in the sound given.*

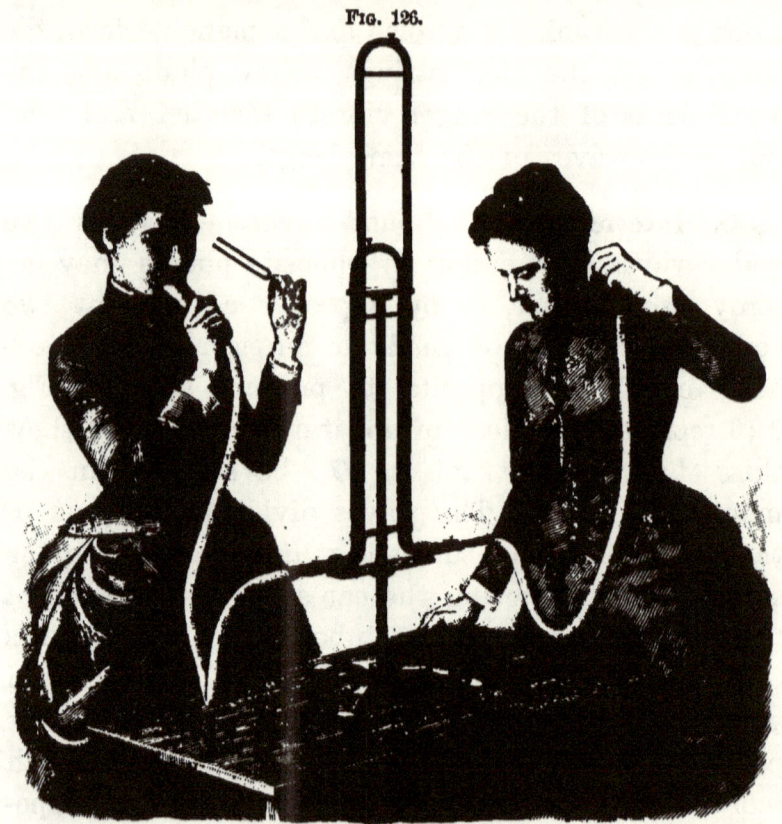

Fig. 126.

Tube for Interference of Sound.

If we strike a tuning-fork and turn it slowly around before the ear, we shall find four points where the interference of the sound-waves causes

* We can not produce complete extinction because, 1. The sound is conducted not only by the inclosed air, but by the solid tube also; 2. There is loss by friction in the longer branch; 3. There is loss by leakage between the tubes that slide against each other.

great weakening. The two prongs swing alternately toward and from each other. When a condensation is produced between the prongs, a rarefaction is produced on their outer sides. Certain lines can be found where these interfere.

If two forks are nearly but not quite in unison, the waves from them are unequal in length. They alternately conjoin and oppose each other, producing "beats." These are often noticed in the sound from a large bell, the opposite sides of which are not quite equally elastic. A pair of mistuned organ-pipes produces a similar effect, and the discord of an inferior piano, or indeed all discord, is due to beats.

7. Vibrations of Cords. — Let ab be a stretched cord made to vibrate. The motion from e to d and

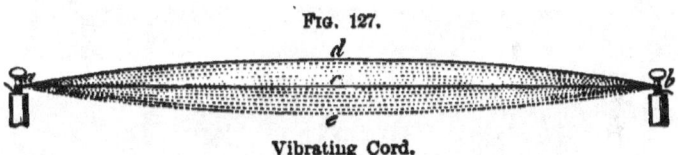

Fig. 127.

Vibrating Cord.

back again is termed a *vibration;* that from e to d, a *half-vibration.* The distance, cd, from the middle to either of the extreme positions is the *amplitude.*

(1.) THE SONOMETER is an instrument used to investigate the laws of vibration of stretched cords. It consists of two cords stretched by weights, P, across fixed bridges, A and B. The movable bridge, D, serves to lengthen or shorten the vibrating part of either cord. Beneath is a resonance box, to which the vibrations are conducted by the bridges. This is the body whose sound is chiefly heard.

172   ACOUSTICS.

(2.) THREE LAWS.—I. *The number of vibrations per second increases as the length of the cord decreases.* By plucking the cord with the finger, or drawing a violin bow across it, make it vibrate, giving the note of the entire string. Place the bridge D at the center of the cord, and the sound will be the *octave* above the former. Thus, by taking one half the length of the cord we double the number of vibrations.—*Examples:* If an entire cord make 20 vibrations per second, one half will make 40, and one

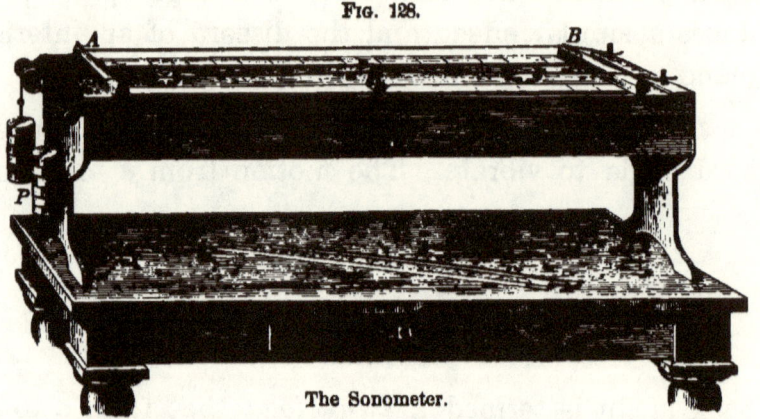

FIG. 128.

The Sonometer.

third, 60.—The violin or guitar player elevates the pitch of a string by moving his finger, thus shortening the vibrating portion.—In the piano, harp, etc., the long and the short strings produce the low and the high notes respectively.

II. *The number of vibrations per second increases as the square root of the tension.* The cord when stretched by 1 lb. gives a certain tone. To double the number of vibrations and obtain the octave requires 4 lbs. Stringed instruments are provided with keys, by which the tension of the cord and the corresponding pitch may be increased or diminished.

III. *The number of vibrations per second decreases as the square root of the weight of the cord increases.* If two strings of the same material be equally stretched, and one have four times the weight of the other, it will vibrate only half as often. In the violin the bass notes are produced by the thick strings. In the piano fine wire is coiled around the heavy strings to increase their weight.

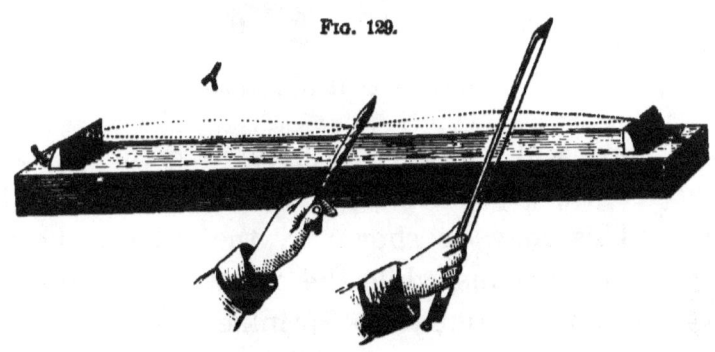

Fig. 129.

Production of Two Segments.

(3.) NODES.—In the experiments just described, the cord is shortened by means of a firm, movable bridge. If, instead, we rest a feather lightly on the string, and draw the bow over one half, the cord will vibrate in two portions and give the octave as before. Remove the feather, and it will continue to vibrate in two parts and to yield the same tone. We can show that the second half vibrates by placing across that portion a little paper rider. On drawing the bow it will be thrown off. Hold the feather so as to separate one third of the string and cause it to vibrate; the remainder of the cord will vibrate in two segments. When the feather is removed, the

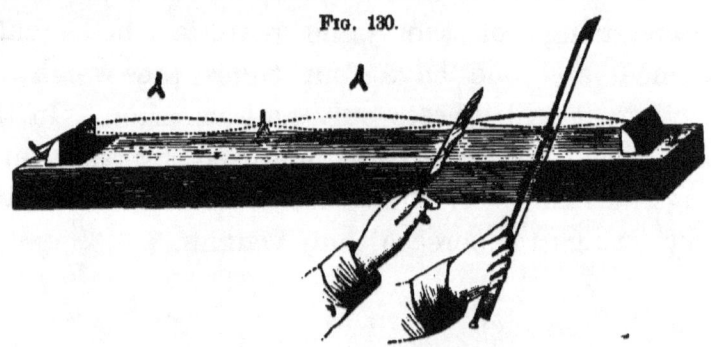

Fig. 130.

Production of Three Segments.

entire cord will vibrate in three different parts of equal length, separated by stationary points called *nodes*. This may be shown by the riders; the one at the node remains, while the others are thrown off.

(4.) ACOUSTIC FIGURES.—Sprinkle fine sand on a metal plate. Place the finger-nail on one edge to stop the vibration at that point, as the feather did in the last experiment, and draw the bow lightly across the opposite edge. The sand will be tossed away from the vibrating parts of the plate and will collect along the nodal

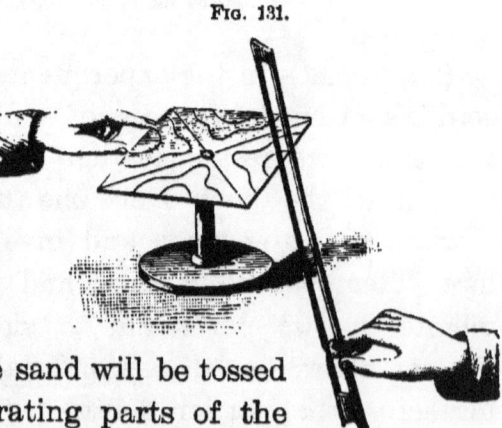

Fig. 131.

Vibration of a Plate.

lines, which divide the large square. It is wonderful to see how the sand will seemingly start into life and dance into line at the touch of the bow. Fig. 132 shows some of the beautiful patterns obtained by Chladni.

FIG. 132.

Chladni's Figures.

(5.) HARMONICS.*— Whenever a cord vibrates, it separates into segments at the same time. Thus we have the full or *fundamental* note of the entire string, and superposed upon it the higher notes produced by the vibrating parts. These are called *overtones* or *harmonics*. The mingling of the two classes of vibrations determines the *quality* of the sound, and enables us to distinguish the music of different instruments.

(6.) NODES OF A BELL.—Let the heavy circle in Fig. 133 represent the circumference of a bell when at rest. Let the hammer strike at a, b, c, or d. At

---

* Press gently but firmly down the notes C, G, and C, in the octave above middle C, on the piano-forte. Without releasing these keys, give to C below middle C a quick, hard blow. The damper will fall, and the sound will stop abruptly. At the same instant a low, soft chord will be heard. This comes from the three strings whose dampers are raised, leaving them free to sound in sympathy with the overtones of the lower C, which sounds are identical with their own.—When a goblet or wine-glass is tapped with a knife-blade, we can distinguish three sounds, the fundamental and two harmonics.

one moment, as the bell vibrates, it forms an oval with *ab*, at the next with *cd*, for its longest diameter. When it strikes its deepest note, the bell vibrates in four segments, with *n, n, n, n*, as the nodal points, whence nodal lines run up from the edge to the crown of the bell. It tends, however, to divide into a greater number of segments, especially if it is very thin, and to produce harmonics. The overtones which accompany the deep tones of the bell are frequently very striking, even in a common call-bell, and often make it hard to determine at once what is its fundamental. Usually they die away sooner than the fundamental.

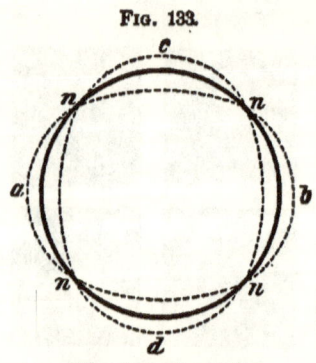

Fig. 133.

Vibration of a Bell.

(7.) NODES OF A SOUNDING-BOARD.—The case of a violin or guitar is composed of thin wooden plates which divide into vibrating segments, separated by nodal lines according to the pitch of the note played. The inclosed air vibrating in unison with these, re-enforces the sound and gives it fullness and richness.

(8.) MUSICAL SCALE.—The lowest tone that can be distinctly perceived as musical by most ears is produced by 32 vibrations per second. This is called C_0. The octave above this is C_1, 64 vibrations; the double octave, C_2, 128 vibrations, etc. If a string be stretched so as to give C_2, the tones of the common musical scale between this and C_3 are obtained from the parts of the string indicated by the following fractions:

C_2	D_2	E_2	F_2	G_2	A_2	B_2	C_3
1	$\tfrac{8}{9}$	$\tfrac{4}{5}$	$\tfrac{3}{4}$	$\tfrac{2}{3}$	$\tfrac{3}{5}$	$\tfrac{8}{15}$	$\tfrac{1}{2}$

As the number of vibrations varies inversely as the length of the cord, we have only to invert these fractions to obtain the relative number of vibrations per second; thus,*

C_2	D_2	E_2	F_2	G_2	A_2	B_2	C_3
1	$\tfrac{9}{8}$	$\tfrac{5}{4}$	$\tfrac{4}{3}$	$\tfrac{3}{2}$	$\tfrac{5}{3}$	$\tfrac{15}{8}$	2
128	144	160	170	192	214	240	256

8. Vibration of Columns of Air.—If a tuning-fork be excited and its prongs be held before the open end of a tube of proper length, the sound will become much louder. If the pitch of the fork is C_4, 512 vibrations, the length of such a tube, open at both ends, is about 13 inches; if open at only one end, $6\tfrac{1}{2}$ inches. A hollow globe of proper size, with an opening on one side, will respond in like manner. Such bodies are called *resonators*.

9. Wind Instruments produce sounds by the vibration of the columns of air which they inclose. An organ-pipe is merely a tube-resonator. The sound-

---

* In this table, "$C_3 = 256$ vibrations" represents the middle C of a piano-forte. This number is purely arbitrary. The so-called "concert-pitch" varies in different countries. The Stuttgart Congress of 1834 fixed the standard tuning-fork—middle A—at 440 vibrations per second; while the Paris Conservatory (1859) gave to middle A 437.5. The pitch agreed upon as an international standard by the Vienna Conference in 1885 for A = 435 vibrations. This was adopted by the Piano Manufacturers' Association of New York and vicinity in 1891. The ratio of the different numbers is identical, whatever the pitch.

178   ACOUSTICS.

waves in organ-pipes are set in motion by either fixed mouth-pieces or vibrating reeds. The air is forced from the bellows into the tube *P*, through the vent *i*, and striking against the thin edge *a*, produces a flutter. The column of air above, thrown into vi-

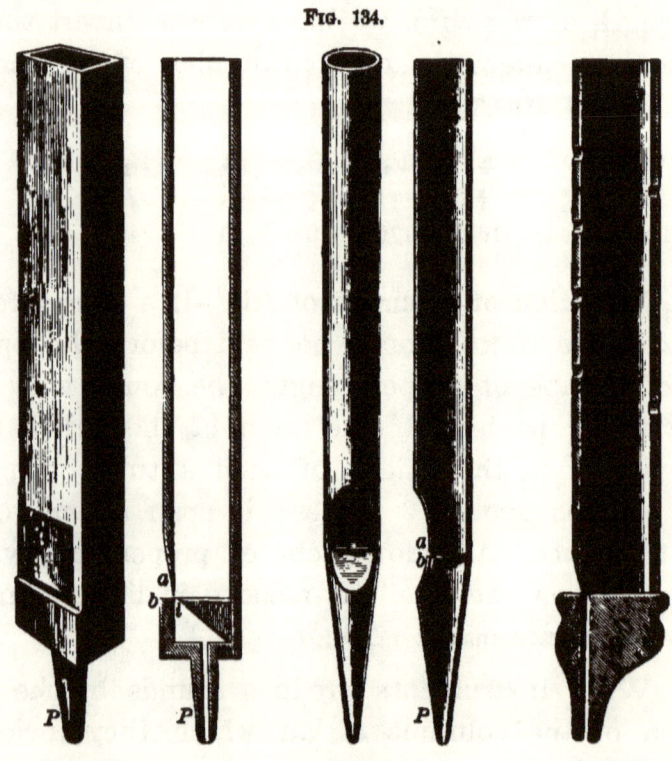

FIG. 134.

Organ-pipes.

bration, re-enforces the sound and gives a full musical tone. The length of the pipe, if open, should be ½ wave length corresponding to the pitch to which it responds; if closed, ¼ wave length. If a tuning-fork which produces this pitch be held at *b* while vibrating, the sound will at once become much stronger.

The air co-vibrates, whatever may be the source of sound, if only the pitch be properly adjusted.

10. Co-vibration.—We have already seen (p. 158) how one tuning-fork may co-vibrate with another through the medium of the air. Vibrations thus produced are often called *sympathetic*, and bodies which thus strengthen sound are said to be *resonant*. Produce a musical tone with the voice near a piano, and a certain wire will seem to select that sound and respond to it. Change the pitch, and the first string will cease, while another replies. If a hundred tuning-forks of different tones are sounding at the foot of an organ-pipe, it will strengthen the sound of the one to which it can reply, and answer that alone. Helmholtz has applied this principle to the construction of the *resonance globe*, an instrument which will respond to a particular harmonic in a compound tone, and strengthen it so as to make it audible.

(1.) SENSITIVE FLAMES.—Flames are sensitive to sound. At an instrumental concert the gas-lights vibrate with certain pulsations of the music. This is noticeable when the pressure of gas is so great that the flame is just on the verge of flaring, and the vibration of the sound-wave is sufficient to "push it over the precipice."*

(2.) SINGING FLAMES.—If we lower a glass tube over a small gas-jet, we soon reach a point where

---

* Prof. Barrett, of Dublin, describes a peculiar jet which is so sensitive that it trembles and cowers at a hiss, like a human being, beats time to the ticking of a watch, and is violently agitated by the rumpling of a silk dress.

the flame leaps spontaneously into song. At first the sound seems remote, but gradually approaches until it bursts into an almost full song. The length of the tube and the size of the jet determine the pitch of the note.* The flame, owing to the friction at the mouth of the pipe, is thrown into vibration. The air vibrates in unison with the jet, and, like that in the organ-pipe, selects the tone corresponding to the length of the tube.

Fig. 135.

Singing Flame.

11. The Phonograph is an instrument for recording and reproducing the vibrations of sound. Its essential features are as follows:

(1.) A metallic cylinder which can be rotated on a screw as axis, so as to secure motion that is sideward as well as rotary.

(2.) A hollow cylinder of wax which fits over the

---

* The jets are easily made by drawing out glass tubing to a fine point over a spirit-lamp.

THE PHONOGRAPH.

metallic cylinder, and may be removed after receiving impressions from a source of sound.

(3.) A mouth-piece into which the speaker vocalizes. At the bottom of this is an elastic disk, which is set into vibration by the voice.

(4.) A lever which is actuated by the disk. At one end of it is a specially prepared needle, which makes indentations upon the rotating cylinder of wax.

The Phonograph

After the line of indentations has been made on the wax, the cylinder is brought back to its first position. On turning it, the needle, pressing on the serrated surface, receives vibratory motion like that which had been given it by the voice. This is received by the disk, and the instrument thus talks out what had been talked into it.

There are usually two mouth-pieces, interchangeable in position, one of which is used in speaking to the phonograph, and the other in giving out what this has to say. Each is specially adapted to the work it has to do. The metallic cylinder is rotated by means of an electric motor. The arm which carries the mouth-pieces is provided with a turning tool for smoothing the wax before this receives the record from the voice.

In Fig. 136, the phonograph is shown ready to talk. A conical speaking-trumpet is fixed upon the mouth-piece, so that the sound may be strengthened by co-vibration. The wax cylinder can be kept any length of time, and be made to speak out its message repeatedly. The phonograph reproduces so accurately the sounds it has received, that even the peculiarities which result from the special quality of the speaker's voice can be recognized.

12. The Human Ear is also an instrument for receiving sound vibrations, which affect the auditory nerve and produce sensation thus at the base of the brain. ("Hygienic Physiology," p. 216.)

(1.) RANGE OF THE EAR.—No definite limit can be assigned to the range through which musical sounds are perceptible. The highest limit has been roughly estimated to be about 38,000, and the lowest, 16, vibrations per second. When the number of impressions on the ear in each second is less than 15 or 16, we become able to perceive them separately. To be musical they must come fast enough to appear to coalesce. From 16 to 38,000 is about eleven octaves. The capacity to hear the higher tones varies in different persons. A sound audible to one may be silence to another. Some ears can not distinguish the squeak of a bat or the chirp of a cricket, while others are acutely sensitive to these shrill sounds. Indeed, the auditory nerve seems generally more alive to the short, quick vibrations than to the long, slow ones. The whirr of a locust is much more noticeable than the sighing of the wind through the trees.*

(2.) THE ABILITY OF THE EAR TO DETECT AND ANALYZE SOUND is wonderful beyond comprehension. Sound-waves chase one another up and down through the air, superposed in entangled pulsations, yet a cylinder not larger than a quill conveys them to the ear, and each string of that wonderful harp selects its appropriate sound, and repeats the music to the soul. Though a thousand instruments be played at once, there is no confusion, but each is heard, and all blend in harmony.†

* To this, however, there are remarkable exceptions. The author knows a lady who is insensible to the higher tones of the voice, but acutely sensitive to the lower ones. Thus, on one occasion, being in a distant room, she did not notice the ringing of the bell announcing dinner, but heard the noise the bell made when returned to its place on the shelf.

† "Is not the ear the most perfect sense? A needle-woman will distinguish by the sound whether it is silk or cotton that is torn. Blind people recognize the age of persons by their voices. An architect, comparing the length of two lines separated from each other, if he estimate within $\frac{1}{30}$, we deem very accurate; but a musician would not be considered very precise who estimated within a quarter of a note (128 ÷ 30 = 4 nearly). In a large orchestra, the leader will distinguish each note of each instrument. We recognize an old-time friend by the sound of his voice, when the other senses utterly fail to recall him. The musician carries in his ear the idea of the musical key and every tone in the scale, though he is constantly hearing a multitude of sounds."

PRACTICAL QUESTIONS.

1. Why can not the rear of a long column of soldiers keep time to the music in front?

2. Three minutes elapse between the flash and the report of a thunderbolt; how far distant is it?

3. Five seconds expire between the flash and the report of a gun; what is its distance?

4. Suppose a speaking-tube should connect two villages ten miles apart; how long would it take the sound to travel?

5. The report of a pistol-shot was returned to the ear from the face of a cliff in four seconds; what was the distance?

6. What is the cause of the difference between the voice of man and woman? A base and a tenor voice?

7. What is the number of vibrations per second necessary to produce the fifth tone of the scale of C_3?

8. What is the length of each sound-wave in that tone when the temperature is at zero?

9. What is the number of vibrations in the fourth tone above C_2?

10. If a meteor were to explode at a height of 60 miles, would it be possible for its sound to be heard at sea-level?

11. A stone is let fall into a well, and in four seconds is heard to strike the bottom; how deep is the well?

12. What time would be required for a sound to travel five miles in the still water of a lake?

13. Does sound travel faster at the foot than at the top of a mountain?

14. Why is an echo weaker than the original sound?

15. Why is it so fatiguing to talk through a speaking-trumpet?

16. Why will the report of a cannon fired in a valley be heard on the top of a neighboring mountain, better than one fired on the top of a mountain will be heard in the valley?

17. Why do our footsteps in unfurnished dwellings sound so startlingly distinct?

18. Why do the echoes of an empty church disappear when the audience assemble?

19. What is the object of the sounding-board of a piano?

20. During some experiments, Tyndall found that a certain sound would pass through twelve folds of a dry silk handkerchief, but would be stopped by a single fold of a wet one. Explain.

21. What is the cause of the musical murmur often heard near telegraph lines?

22. Why will a variation in the quantity of water in a goblet, when this is made to sound, cause a difference in the tone produced by its vibration?

23. At what rate (in meters) will sound move through air at sea-level, the temperature being 20° C.?

SUMMARY.

SOUND is produced by vibrations. These are transmitted in waves through the air (60° F.) at sea-level at the rate of 1,120 ft. per second; through water four times, and through iron fifteen times as fast. In general, the velocity depends on the relation between the density and the elasticity of the medium; and the intensity is proportional to the square of the amplitude of the vibrations. Sound, like light, may be reflected and refracted to a focus. Echoes* are produced by the reflection of sound from smooth surfaces, not less than 112 ft. (about 33 meters) distant. Rapidly-repeated vibrations make a continuous sound; regular and rapid vibrations produce music; irregular ones cause a noise.

The pitch of a sound depends on the rapidity of the vibrations. The number of waves, and their consequent length in a given sound, is found by means of the siren. Unison is produced by identical wave-motions. Any number of sound-waves may traverse the air, as any number of water-waves may the surface of the sea, without losing their individuality. The motion of each molecule of air is the algebraic sum of the several motions it receives. Two systems of waves may therefore destroy or strengthen each other, according as they meet in oppo-

---

* Several acoustic phenomena have become of historical interest. (1.) Near Syracuse, Sicily, is a cave known as the Ear of Dionysius. A whisper at the farther end of the cavern is easily heard by a person at the entrance, though the distance is 200 ft. Tradition says that the Tyrant of Syracuse used this as a dungeon, and was thus enabled tó listen to the conversation of his unfortunate prisoners. (2.) On the banks of the Nile, near Thebes, is a statue 47 ft. high, and extending 7 ft. below the ground. It is called the Vocal Memnon. Ancient writers tell us that about sunrise each morning, there issued from this gigantic monolith a musical sound resembling the breaking of a harp-string. It is now believed that this was produced by friction due to unequal expansion of different parts under the morning sun. (3.) Near Mount Sinai, in Arabia, remarkable sounds are produced by the sand falling down a declivity. The sand, which is very white, fine, and dry, lies at such an angle as to be easily set in motion by any cause, such as scraping away a little at the foot of the slope. The sand then rolls down with a sluggish motion, causing at first a low moan, that gradually swells to a roar like thunder, and finally dies away as the motion ceases.

site or in similar phases. Interference is the mutual weakening of two systems of waves which meet in opposite phases. Beats are the effect produced by two musical sounds of nearly the same pitch, which alternately interfere and coalesce. The vibrations of a cord produce a musical sound, which is re-enforced by a sounding-board. The rate of vibration and consequent pitch depends on the length, the tension, and the weight of the cord. A sounding body vibrates not only as a whole, but also in segments. Its vibration as a whole produces the fundamental tone, and the additional vibration in segments gives rise to the overtones. These together form either a complete or an interrupted harmonic series. The quality of the compound sound depends on the number, orders, relative intensities, and mode of combination of the overtones into which it can be resolved. The various notes in the musical scale are determined by fixed portions of the length of the cord. The music of a wind instrument is produced by vibrating columns of air. Resonance is a sympathetic vibration caused by one sonorous body acting on another, through a conducting medium, as seen in the resonance globe, etc. The voice is a reed instrument, with its vibrating cords and resonant cavity. The ear collects the sound-waves. It consists of the outer ear, the drum, and the labyrinth. The auditory nerve transmits to the brain the motions produced in the ear by sound-waves.

HISTORICAL SKETCH.

THE ancients knew that without air we should be plunged in eternal silence. "What is the sound of the voice," cried Seneca, "but the concussion of the air by the shock of the tongue? What sound could be heard except by the elasticity of the aërial fluid? The noise of horns, trumpets, hydraulic organs, is not that explained by the elastic force of the air?" Pythagoras, who lived in the sixth century before Christ, conceived that the celestial spheres are separated from each other by intervals corresponding with the relative lengths of strings arranged to produce harmonious tones. In his musical investigations he

HISTORICAL SKETCH. 187

used a monochord, the original of the sonometer now employed by physicists, and wished that instrument to be engraved on his tomb. Pythagoras held that the musical intervals depend on mathematics; while his great rival, Aristoxenes, claimed that they should be tested by the ear alone. The theories of these two philosophers long divided the attention of the scientific world. The former considered the subject from the stand-point of Physics, the latter from that of Physiology.

Many centuries elapsed before any marked advance was made. Galileo called attention to the sonorous waves traversing the surface of a glass of water, when the glass is made to vibrate. He gave an accurate explanation of the phenomena of resonance, and referred to the fact that every pendulum has a fixed oscillation period of its own; that a succession of properly timed small impulses may throw a heavy pendulum into vibration, and that this may communicate vibration to a second pendulum of the same vibration period. Galileo also described the first experiment involving the direct determination of a vibration ratio for a known musical interval. He related that he was one day engaged in scraping a brass plate with an iron chisel, in order to remove some spots from it, and noticed that the passage of the chisel across the plate was sometimes accompanied by a shrill whistling sound. On looking closely at the plate, he found that the chisel had left on its surface a long row of indentations parallel to each other and separated by exactly equal intervals. This occurred only when a sound was heard. It was found that a rapid passage of it gave rise to a more acute sound, a slower passage to a graver sound, and that in the former case the indentations were closer together. After many trials, two sets of markings were obtained, which corresponded to a pair of tones making an exact fifth with each other. The indentations were 30 and 45, respectively, to a given length. Galileo's inference from this was exactly what we now accept as true.

The present century has witnessed a more complete demonstration of the laws of the vibrations of cords and the general principles of sound. In 1822, Arago, Gay-Lussac, and others decided the velocity of sound to be 337 *meters* at 10° C. Savart invented a toothed wheel by which he determined the number of vibrations in a given sound; Latour invented the siren, which gave still more accurate results; Colladon and Sturm, by a series of experiments at Lake Geneva, found the velocity of sound in water; Helmholtz made known the laws of harmonics; Lissajous, by means of a mirror attached to the vibrating body, threw the vibrations on a screen in a series of curves, and so rendered them visible; while Tyndall has investigated the causes modifying the propagation of sound, as acoustic clouds, fogs, etc., and popularized the whole subject of acoustics.

VII.

ON LIGHT.

THE sunbeam comes to the earth as simply motion of ether-waves, yet it is the grand source of beauty and power. Its heat, light, and chemical energy work every-where the wonder of life and motion. In the growing plant, the burning coal, the flying bird. the glaring lightning, the blooming flower, the rushing engine, the roaring cataract, the pattering rain—we see only varied manifestations of this one protean energy which we receive from the sun.

ANALYSIS OF LIGHT.

OPTICS, OR THE SCIENCE OF LIGHT.

1. **PRODUCTION AND PROPAGATION OF LIGHT.**
   1. Definitions.
   2. Visual Angle.
   3. Laws of Light.
   4. Velocity of Light.
   5. Theory of Light.

2. **REFLECTION OF LIGHT.**
   1. Definition and Law.
   2. Action of Rough and Polished Surfaces.
   3. Mirrors.
      - (1.) *Plane.*
      - (2.) *Concave.*
      - (3.) *Convex.*

3. **REFRACTION OF LIGHT.**
   1. Definition, and Illustrations.
   2. Laws of Refraction, and Illustrations.
   3. Lenses.
      - (1.) *Convex.*
      - (2.) *Concave.*
   4. Spherical Aberration.
   5. Total Reflection.
   6. Mirage.

4. **DECOMPOSITION OF LIGHT.**
   1. The Prismatic Spectrum.
   2. Solar Energy.
   3. Properties of the Spectrum.
   4. Interruptions in the Spectrum.
   5. The Spectroscope.
   6. Three Kinds of Spectra.
   7. Color.
   8. Complementary Colors.
   9. The Rainbow.
   10. Chromatic Aberration.
   11. Polarization.
       - (1.) *Definition.*
       - (2.) *By Double Refraction.*
       - (3.) *By Reflection.*
       - (4.) *The Polariscope.*

5. **OPTICAL INSTRUMENTS.**
   1. Microscope.
   2. Telescope.
   3. Opera-glass.
   4. Projecting Lantern.
   5. Camera.
   6. The Eye.
   7. The Stereoscope.

OPTICS, OR THE SCIENCE OF LIGHT.

I. PRODUCTION AND TRANSMISSION OF LIGHT.

1. Definitions.—A *luminous* body is one that emits light. A *medium* is any substance through which light passes. A *transparent** body is one that obstructs light so little that we can see objects through it. A *translucent* body is one that lets some light pass, but not enough to render objects visible through it. An *opaque* body is one that does not transmit light. A *ray of light* is a single line of light. A *pencil* or *beam of light* is a collection of rays, which may be *parallel, diverging,* or *converging;* it may be traced in a dark room into which a sunbeam is admitted by the floating particles of dust which reflect the light to the eye.

2. The Visual Angle is the angle formed at the eye by rays coming from the extremities of an ob-

---

* The terms transparent and opaque are relative. No substance is perfectly transparent, or entirely opaque. Glass obstructs some light. According to Miller, 7 ft. of the clearest water will arrest one half the light which falls upon it. While Young asserts that the beam of the setting sun, passing through 200 miles of air, loses $\frac{199}{200}$ of its force. On the other hand, gold, beaten into leaf, becomes translucent, transmitting green light; and scraped horn is semi-transparent.

ject. The angle AOB, is the angle of vision subtended by the object AB. The size of this angle varies with the distance of the body. AB and $A'B'$ are of the same length, and yet the angle $A'OB'$ is

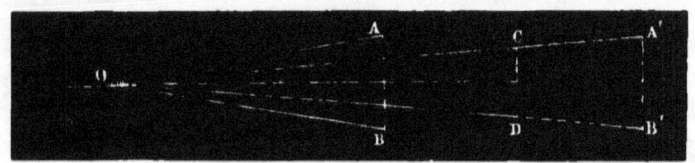

FIG. 137.

Variation of Visual Angle with Distance.

smaller than AOB, and hence $A'B'$ will seem shorter than AB. The distance and the apparent size of objects are intimately connected, since by experience we have learned to associate them. Knowing the distance of an object, we immediately estimate its size from the visual angle.*

3. Laws of Light.—1. Light passes off from a luminous body equally in every direction. 2. Light travels through a uniform medium in straight lines. 3. The intensity of light decreases as the square of the distance increases.

4. The Velocity of Light has been determined in various ways. The following was the first method: The planet Jupiter has five moons. As these revolve around the planet, they are eclipsed at regular intervals. In Fig. 138, let J represent Jupiter, e one of the moons, S the sun, and T and t different positions of the earth in its orbit around the sun. When the

---

* We can *vary the apparent size of any body at which we are looking by increasing or diminishing this angle*—a principle that will be found of great importance in the formation of images by mirrors and lenses.

earth is at T, the eclipse occurs 16 min. and 36 sec. earlier than at t. That interval of time is required

Fig. 138.

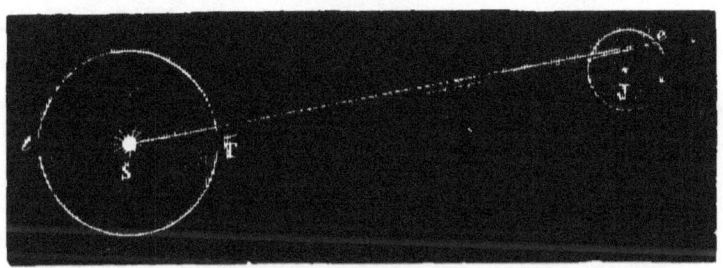

The Sun, Earth, and Jupiter.

for the light to travel across the earth's orbit, giving a velocity of about 186,000 miles per second.*

5. Undulatory Theory of Light.—To account for the phenomena presented by light a substance is assumed to exist which pervades all space. To this substance the name *luminiferous ether*, or, more shortly, *ether*, has been given. In it the heavenly bodies are immersed, and with it the pores of all substances are filled, so that matter may be spoken of as being "ether-soaked." The best attainable vacuum still contains this substance. A luminous body sets in motion waves of ether, which go off in every direction. They move at the rate of 186,000 miles per second, and, breaking upon the eye, give the impression of light. In the wave-motion of light, the vibrations are *transverse* (crosswise) to the direction of propagation.†

---

* This rate is so great that for all distances on the earth it is instantaneous. A sunbeam would girt the globe quicker than we can wink, if its path could be appropriately curved.

† Thus, if we suppose a star directly overhead and a ray of light coming down to us, we should conceive that some of the particles which com-

II. REFLECTION OF LIGHT.

1. Definition.—Light falling on a surface is divided into two portions. One enters the body; the other is reflected* according to the familiar law of Motion and of Sound: The angle of incidence is equal to that of reflection.

2. Action of Rough and Polished Surfaces.— When the surface is rough, the numerous little elevations scatter the reflected rays in every direction, forming *diffused* light. Such a body can be seen from any point. When the surface is polished, the rays are uniformly reflected in particular directions, and may bring to us the images of other objects. We thus see non-luminous objects by irregularly-reflected (diffused) light, and images of objects by regularly-reflected light. †

3. Mirrors.—All highly-reflecting surfaces are mirrors. These are of three kinds—*plane, concave*, and *convex*. The first has a flat surface; the second, one

pose the waves are vibrating E. and W., others N. and S., and others toward all other possible points of the compass in succession.

* The amount of light reflected varies with the angle at which light falls. Thus, if we look at the images of objects in still water, we notice that those near us are not so distinct as those on the opposite bank. The rays from the latter striking the water more obliquely, are more perfectly reflected to the eye.—Fill any dark-colored pail with water tinted with bluing or red ink. The color will be quite invisible to a spectator at a little distance. Now insert in the water a plate. This will reflect the transmitted light and reveal the hue of the water.

† The most perfectly polished substance, however, diffuses some light— enough to enable us to trace its surface; were it not so, we should not be aware of its existence. The deception of a large plate-glass mirror is often nearly complete; but dust or vapor, increasing the irregular reflection, will bring its surface to view.

REFLECTION OF LIGHT.

like the inner surface of a hollow globe; the third, like part of its outer surface. The general principle of mirrors is that *the image is seen in the direction of the reflected ray as it enters the eye.*

(1.) PLANE MIRRORS.—Rays of light retain their relative direction after reflection from a plane surface.* While standing before a plane mirror, one sees his image erect and of the same size as himself. It is, however, reversed right and left.

Why the image is as far behind the mirror as the object is in front. Let AB be an arrow held in front of the mirror MN. Rays of light from the point A striking upon the mirror at C, are reflected, and enter the eye as if they came from a. Rays from B seem to come from b.

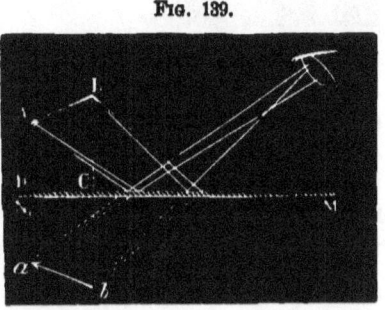

FIG. 139.

Since the image is seen in the direction of the reflected rays, it appears at ab, a point which can easily be proved to be as far behind MN as the arrow is in front of it. Such an image is called a

---

* The perpendiculars are not given in the figures of the book, as *the pupil at recitation should draw all the cuts on the blackboard, erect the perpendiculars, and demonstrate the location of the reflected ray.* It will aid in drawing the perpendicular to a convex or concave surface, to remember that it is a radius of the sphere of which the mirror forms a part. A book held in various positions before a looking-glass illustrates the action of plane mirrors. A beam of light admitted into a dark room and reflected from a mirror will show that the angles of incidence and reflection are in the same plane. Many of the grotesque effects of concave and convex mirrors may be seen on the inner and outer surfaces of a bright spoon, call-bell, or metal cup (see "Mayer & Barnard's Light" for inexpensive experiments).

virtual one, as it has no real existence apart from the observer's eye.

Why we can see several images of an object in a mirror. Metallic mirrors form only a single image.

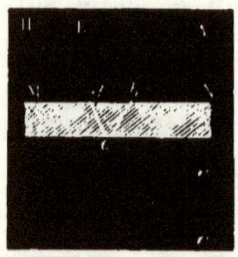

Fig. 140.

If, however, we look obliquely at the image of a candle in a looking-glass, we shall see several images, the first feeble, the next bright, and the others diminishing in intensity. The ray from A is in part reflected to the eye from the glass at b, and gives rise to the image a; the remainder passes on and is reflected from the metallic surface at c, and coming to the eye forms a second image a'. The ray cd, when leaving the glass at d, loses a part, which is reflected back to form a third image. This ray in turn is divided to form a fourth, and so on.

If two mirrors are arranged as in Fig. 141, three images of a candle may be seen. (Let the pupil trace the formation of each by the diagram of Fig. 142.) To vary the experiment, hold the mirrors together like the covers of a book placed on end, and put the candle between them on the table, opening and shutting the mirror-cover so as to vary the angle; or hold the mirrors parallel to each other with the light between them. When the mirrors are inclined at 90°, three images are formed; at 60°, five images; and at 45°, seven images. As the angle increases, the number diminishes. The images are upon the circumference of a circle whose center is on a line in which the reflecting surfaces would intersect if produced. Where the mirrors are parallel the

FIG. 141.

Multiple Reflection.

images are in a straight line. They become dimmer as they recede, light being lost at each reflection.—The *Kaleidoscope* contains three mirrors set at an angle of 60°. Small bits of colored glass at one end reflect to the eye at the other multiple images which change in varying patterns as the tube is revolved.

FIG. 142.

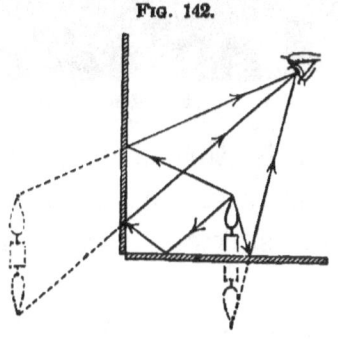

Images seen in water are symmetrical, but inverted. The reason of this can be understood by holding an object in front of a horizontal looking-glass and noticing the angle at which the rays must strike the surface in order to be reflected to the eye.

When the moon is high in the heavens, we see the image in the water at only one spot, while the rest of the surface appears dark. The light falls upon all parts, but each ray is reflected from only one point at the proper angle to reach the eye. Each observer sees the image at a different place. When the surface of the water is ruffled, a tremulous line of light is reflected from the side of each tiny wave that is turned toward us. As every little billow rises, it flashes a gleam of light to our eyes, and then sinking, comes up beyond, to reflect another ray.

Fig. 148.

Reflection of Light.

(2.) A CONCAVE MIRROR tends to collect the rays of light to a focus. In Fig. 144, C is the *center of curvature*, *i.e.*, the center of the hollow sphere of which the mirror is a part; V is the *vertex*, or middle of the mirror; F is the *principal focus;* it is half-way between C and V. Any ray which passes through C is an *axis;* it is called the *principal* axis if it pass also through V, otherwise it is a *secondary axis*. All axial rays are reflected back upon their own paths. All rays parallel to the principal axis cross at the principal focus after reflection, and conversely all rays which pass through the principal focus will be reflected parallel to the principal axis.* An image is *real* if the rays after reflection cross before reaching the eye; it will appear to be at the crossing point. Otherwise, the image is *virtual*.

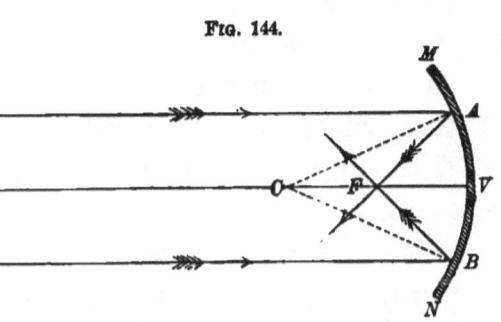

Fig. 144.

Parallel Rays Reflected to the Focus.

Images formed by Concave Mirrors.—In a dark room place a candle (PQ, Fig. 145) in front of a concave mirror at some distance beyond its center of curvature. A *small inverted image* of it will appear to be suspended in mid-air near its focus. It is easy to determine the position of this image. From P draw an axial ray through C; it will be reflected back on its own path, hence the image of P must be

---

* These statements are approximately true only for mirrors of slight curvature, where the angle MCN, or *angular aperture*, does not exceed 8° or 10°. When greater, the rays reflected near the edge of the mirror meet the *principal axis* VC, nearer the mirror than F. This is called the *aberration* of the mirror. The reflected rays will then cross at points in a curved surface called a *caustic*. A section of such a curve can be seen when the light of a candle is reflected from the inside of a cup partly full of milk. All of these phenomena can be proved mathematically to be necessary consequences of the one law, that the angles of incidence and reflection are equal.

on this line. From P draw also a ray, Pa, parallel to the principal axis; after reflection it will pass through the focus, F, and cross the secondary axis at P', which is hence the position of the image of P. In like manner we may determine Q'. If a piece of thin white paper or roughened glass be put at $P'Q'$, the light will seem to come from it since the rays cross here.

Bring the candle closer to the mirror. The image will grow larger and move from F toward C. When the candle reaches C, the image will fall upon it and

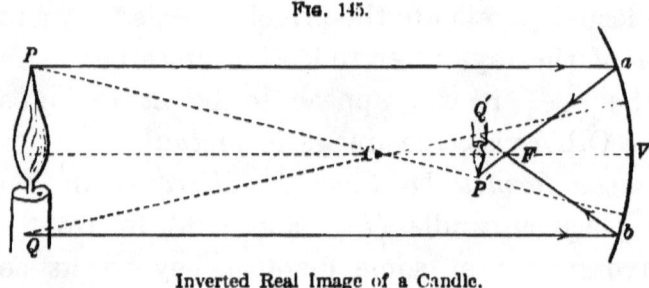

Fig. 145.

Inverted Real Image of a Candle.

just cover it. When it reaches $P'Q'$, the image will have receded to PQ, and in every case *the ratio of the lengths of candle and image will be the same as the ratio of their distances from C.* P and P' are called *conjugate points;* for they are so related that an object placed at one of them will be imaged at the other.

If the candle be moved toward F, the image increases in size and recedes from C. At F the image vanishes. Passing F the image again appears, but now as if it were *behind* the mirror, *larger* than the object *and erect.* Let the student trace the rays, as shown in Fig. 146, and satisfy himself that they can never cross after reflection. The image is hence *virtual;* it can not be caught on a screen; but its apparent length is as much *greater* than that of the candle as its apparent distance from C is greater. Moreover, it appears erect, and not inverted like the real image of the more distant candle.

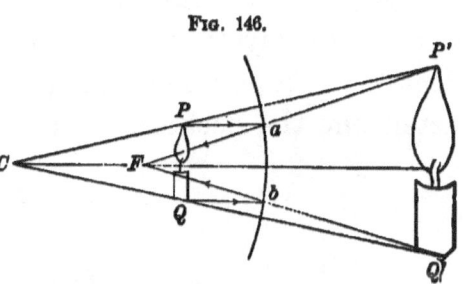

Fig. 146.

Virtual Image in a Concave Mirror.

(3.) CONVEX MIRRORS.—Let the position of a candle be varied in front of a convex mirror. It will be found that the image is always *virtual, erect,* and

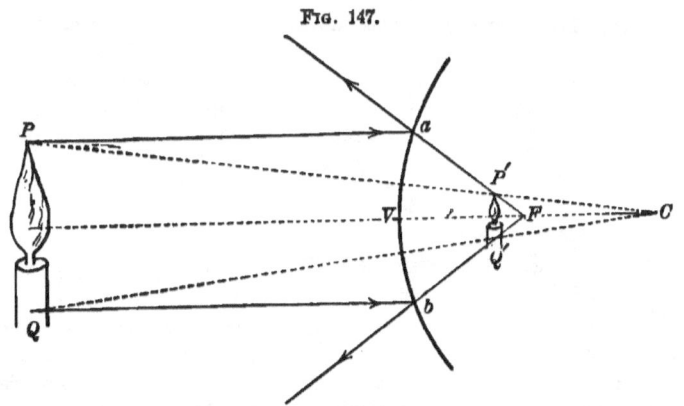

Fig. 147.

Virtual Image in a Convex Mirror.

smaller than the candle. Parallel rays are made to *diverge* after reflection, as if they had come from a point within the sphere, half-way between its surface and center. The image of *P* is at the crossing point of the axial ray from *P* and the backward prolongation of the ray from *P* which was parallel before reflection. The student can easily trace the rays and determine the position of the image.

III. REFRACTION OF LIGHT.

1. Definition.—When a ray of light passes obliquely from one medium to another of different density, it is *refracted* or bent out of its course.—*Examples:* A spoon in clear tea appears bent.—An oar dipping in still water seems broken at the point where it enters the water.*—Put a cent in a bowl. Standing where you can not see the coin, let another

Fig. 148.

Apparent breaking of a Stick in Water.

* In Fig. 148 is shown a stick sunk till the end is at the bottom of the water. Rays of light from this end are bent as they emerge from the liquid and reach the eye as if they had come from a point considerably higher. The entire bottom, therefore, seems lifted up. Hence, water is always deeper than it appears. Look obliquely into a pail of water, then place your finger on the outside where the bottom seems to be; you will be surprised to find the real bottom is several inches below.—Fill a glass dish with water, and, darkening the windows, let a sunbeam fall upon the surface. The ray will bend as it enters. Dust scattered through the air will make the beam distinct.

person pour water into the vessel, when the coin will be lifted into view. To understand the apparent change of position, remember that *the object is seen in the direction of the refracted ray as it enters the eye.* Let L, Fig. 149, be a body beneath the water. A ray, LA, coming to the surface, is bent away from the vertical, LK, and strikes the eye as if it came from L. The object will therefore apparently be elevated above its true place.

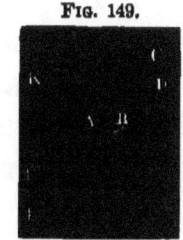

Fig. 149.

2. **Laws of Refraction.**—From any point, A, Fig. 150, let a beam of light, AB, pass through air and meet a denser transparent medium at B, such as water or glass. At this point let a line, DE, be drawn perpendicular to the surface. Then some of the light will be reflected at B, the angle of reflection DBF being equal to the angle of incidence, DBA. A little of it will be absorbed and changed into heat. The rest will be transmitted, but its direction changed to BC. This apparent breaking of the ray is called *refraction*, and the angle EBC, which is less than DBA, is the *angle of refraction*. If the source of light were at C, its direction on emerging at B

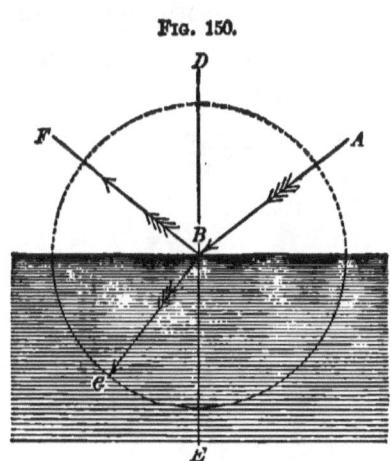

Fig. 150.

Reflection and Refraction.

would be *BA*. The angle of refraction now is *DBA* and the angle of incidence is *EBC*. Hence,

I. In passing into a denser medium, the ray is bent *toward* the perpendicular.

II. In passing into a rarer medium, the ray is bent *from* the perpendicular.*

ILLUSTRATIONS.—*Path of rays through a window-glass.*—When a ray enters a window-glass, it is refracted toward the perpendicular (1st law), and, on leaving, is refracted equally from the perpendicular (2d law). The general direction of objects is therefore unchanged. A poor quality of glass produces distortion by its unequal density and uneven surface.

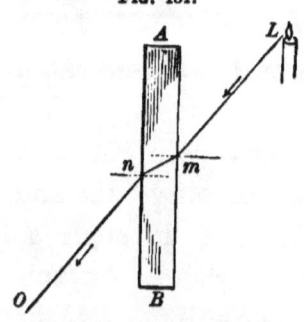

Fig. 151.
Passage of a Ray through Window-glass.

Path of rays through a prism.—A ray of light, on entering and on leaving a glass prism, is refracted. The inclination of the sides causes the ray to be *bent twice in the same direction.* The candle, *L*, will therefore appear to be in the direction of *r*.

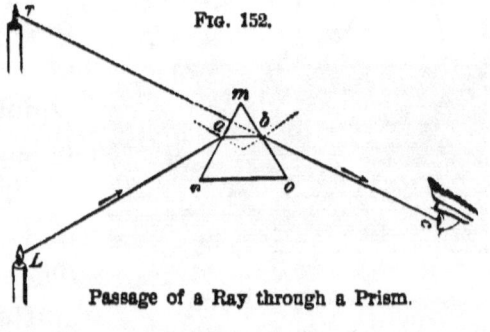

Fig. 152.
Passage of a Ray through a Prism.

---

* Both the incident and the refracted ray lie in the same plane as the normal (perpendicular). The ratio between the sines of the angles of incidence and refraction is termed the *index of refraction*. It varies with the media.—*Example:* From air to water it is ¾ and from air to glass ⅔.

3. Lenses.—A lens is a transparent body, with at least one curved surface. There are two general classes of lenses, *concave* and *convex.** (See Fig. 153.)

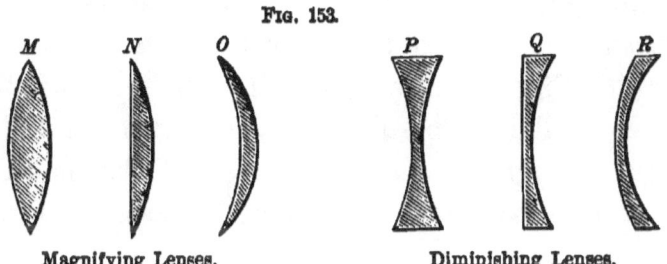

Fig. 153.

Magnifying Lenses. Diminishing Lenses.

(1.) THE BI-CONVEX LENS has two convex surfaces. Its action on light is like that of a concave mirror. A ray, X, striking perpendicularly, is not refracted.

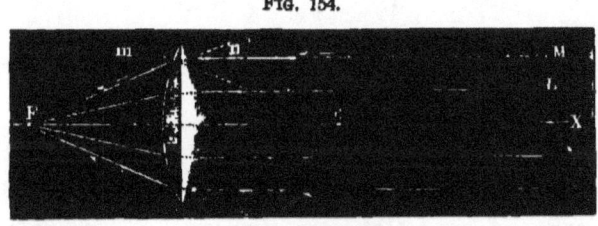

Fig. 154.

Bi-convex Lens.

The parallel rays, M, L, etc., are refracted both on entering and on leaving the lens, and are converged at F, the principal *focus*.† If a luminous point be placed at F, its rays will emerge parallel.

---

* Forms of lenses: M, double-convex; N, plano-convex; O, meniscus (crescent); P, double-concave; Q, plano-concave; R, concavo-convex. The first three are styled *magnifiers*, and the second, *diminishers*.

† The convex lens is sometimes termed a *burning-glass*, from the fact that, like a concave mirror, it collects and brings to a focus the rays of the sun; and combustible substances placed at this focus may be *burned*. Even glass globes of water, such as are used for gold-fishes or in the windows of drug stores, may fire adjacent objects.

Construction of Images.—There is a point, called the *optical center* of the lens, through which the passing ray does not change its general direction after emergence. These are called *axial* rays. The *principal* axis passes not only through the optical center (C, Fig. 155), but also through the principal focus, F. All rays parallel to it are so refracted as to pass through F. Let a candle, PQ, be placed in front of a bi-convex lens,* at a distance of ten or

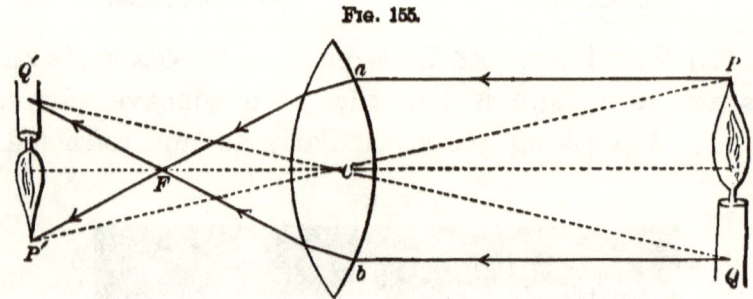

Fig. 155.

Formation of an Image with a Bi-convex Lens.

twelve feet. An axial ray, PC, continues its path unchanged. A parallel ray, Pa, will after refraction pass through F. Where this cuts the axial ray at P', the image of P is found. In like manner Q' is found as the point conjugate to Q. The image, P'Q', is *real*, and may be caught on a screen. It is *inverted*, and as much *smaller* than the object as its distance from the optical center is less.

As the candle is made to approach the lens on one side, the image recedes on the other. When

---

* An ordinary magnifying hand-glass, such as is often used in looking at photographs, or even a spectacle-lens used by an aged person in reading, will be sufficient for these experiments.

brought nearly to F, the image on the other side grows very large and distant. When it arrives *within* the focal distance, FC, the image suddenly appears on the same side with the object, *erect*, and as much larger as its apparent distance from C is greater.

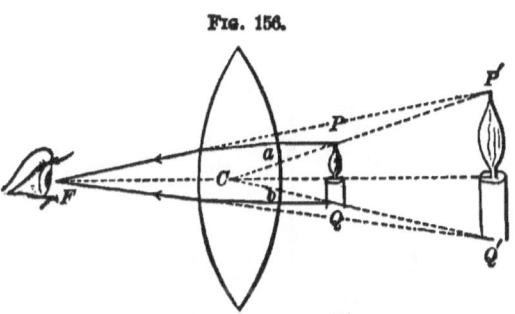

Virtual Image with Convex Lens.

This image is *virtual*. The student can determine this by tracing the rays in Fig. 156.

(2.) THE BI-CONCAVE LENS has two concave surfaces. Its action on light is like that of a convex mirror. Thus, diverging rays from L (Fig. 157) are

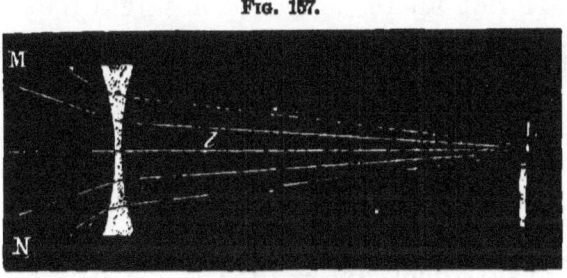

Bi-concave Lens.

rendered more diverging, and, to an eye which receives the rays MN, the candle would seem to be at l, where the image is seen.*

---

* Unscrew the eye-piece from an opera-glass; it serves well for experiments with concave lenses. Unscrew the glass at the other end; it serves for those with convex lenses.

The image formed by a concave lens, like that of a convex mirror, is *virtual, erect*, and *diminished* in

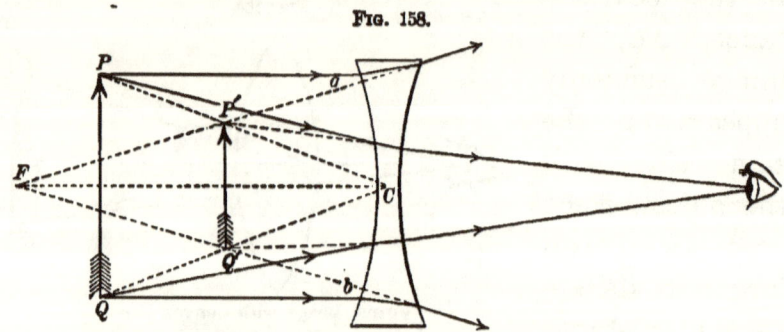

Fig. 158.

Formation of an Image with a Concave Lens.

size (Fig. 158). Let the student determine this by tracing the rays in Fig. 158, in which the arrow PQ is the object and $P'Q'$ the image.*

4. Spherical Aberration of Lenses.—Parallel rays falling on a lens whose surface is like that of a sphere are not all refracted to a single focus. Those which pass through marginal parts of the lens are collected to a focus nearer than that to which the central rays are collected. With a single lens, therefore, it is not easy to secure perfect distinctness of image.

5. Total Reflection.—In passing from a dense into a rare medium, the angle of refraction is greater than the angle of incidence. It can not exceed 90°, for then the ray would cease to emerge. The angle

---

* Remember that parallel rays, Pa and Qb, become divergent after refraction, as if they had come from a focus, F, on the same side of the lens. The images of P and Q must hence be found on the *backward* prolongations of these emergent rays. Axial rays are drawn from P and Q to C.

of incidence for which the emergent ray would make an angle of 90° with the perpendicular is called the *critical angle*. The light is then totally reflected in the dense medium as if its surface were the most perfect of mirrors. When we look obliquely into a pond, we can not see the bottom, because the rays of light from below are reflected downward at the surface of the water. Hold a glass of water above the level of the eye, and the upper part will gleam like burnished silver.* Thus the internal surface of a transparent body becomes a mirror.

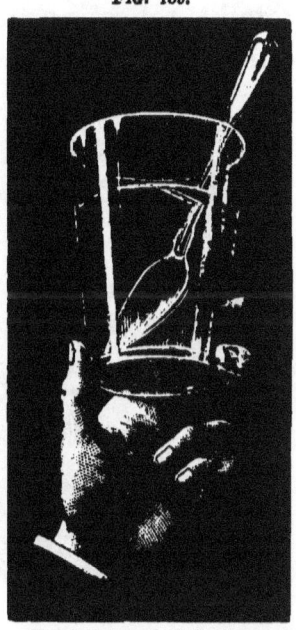

Fig. 159.

Total Internal Reflection.

6. Mirage.—Over the heated deserts of Arabia and Africa the traveler sometimes sees a shimmering expanse, as if a quiet lake were in the distance, in which the scattered trees are mirrored upside down. The layer of air close to the uniformly heated sand is less dense than the cooler air above. A ray coming obliquely downward from a tree-top may be so bent from its first direction by passing through

---

* Place a bright spoon in the glass and notice its image reflected from the surface of the water. Turn the spoon about in the glass and, changing the angle of observation, notice the effect. The real handle may apparently be attached to the image in the water. The spoon will soon be covered with bubbles of air shining, like pearls, from total reflection. This shows also the presence of air in water and the adhesion of gases to solids.

these different media as to be sent obliquely upward to the eye. The low warm layer of air acts like a totally reflecting mirror, and inverted images are dimly seen amid the bright light along the horizon.

Fig. 160.

Mirage.

In Fig. 160, rays of light from a clump of trees are refracted more and more until finally they are bent upward from a layer at *a*, and enter the eye of the Arab as if they came from the surface of a quiet lake.

IV. COMPOSITION OF LIGHT.

1. The Prismatic Spectrum.—When a sunbeam is received through a narrow slit and transmitted through a prism, properly placed, the ray is not only bent from its course, but is also spread out into a band of rainbow colors—the solar spectrum. This includes a multitude of tints grading imperceptibly

COMPOSITION OF LIGHT. 211

from one to another. The most prominent are *violet, indigo, blue, green, yellow, orange, red.** If we receive the spectrum on a concave mirror or pass it through a convex lens appropriately adjusted in position, these colors may be recombined so as to form a white band. We therefore conclude that

Fig. 161.

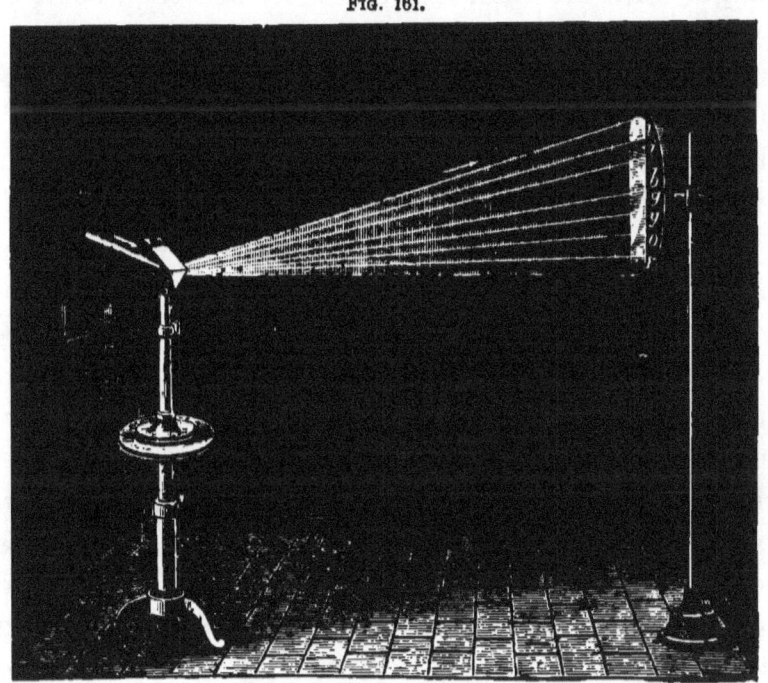

The Prismatic Spectrum.

white light is made up of these many tints. Because each has its own separate index of refraction (see p. 204, foot-note) when passing through the same prism, this refracts them unequally. The deviation of the violet is the greatest, and that of the red is the least, for the visible rays of the spectrum.

* Notice that the initial letters spell the mnemonic word, *Vib-gy-or.*

2. Solar Energy.—What we receive from the Sun is called SOLAR ENERGY. It reaches us in tiny waves, the longest of which are so minute that 8,000 of them in succession would be required to cover an inch. The shortest that have been measured are about a tenth as long, or $\frac{1}{80000}$ of an inch. The longer ones are manifested largely as heat; some of the intermediate ones as light, and the shortest as chemical energy. All these waves come mixed together in the sunbeam. The prism changes their direction, and arranges them in the order of their wave-lengths.

3. Properties of the Spectrum.—Only a small part of the solar spectrum is perceptible to the eye. Beyond the violet all is dark, but by employing a photographic plate this invisible part of the spectrum may be photographed. By means of Langley's bolometer (page 347), an instrument sensitive to variations of temperature, we may explore the region beyond the red, and show that the spectrum continues also in this direction. Solar energy appears thus to have been separated by the prism into parts having different properties. Part producing the sensation of light, another part lying beyond the red manifesting itself mainly as heat, and still another part beyond the violet distinguished by its chemical effects. This difference in properties is, however, only apparent. There is physically no distinction between the rays occupying different parts of the spectrum except a difference of wave-length. A ray belonging to any part of the visible spectrum shows all three properties.

It affects the eye with the sensation of color, the photographic plate in producing chemical change, and the bolometer in heating the filament of platinum.

4. Interruptions in the Spectrum.—When the spectrum of the sun is carefully examined, it is found that there are numerous breaks in both the visible and invisible parts. Numerous black lines, parallel to the slit that transmits the light, may be detected in the visible part. The more prominent of these have been named. Thus, the A, B, and C lines are in the red; the D line in the yellow; the E line in the green; the F line in the blue; the G and H lines in the violet. The interruptions in the invisible portion are far broader, becoming bands rather than narrow lines.

5. The Spectroscope is an instrument used for the production and examination of spectra. It consists of one or more prisms for producing the spectrum, and a telescope for examining it. From the source of light, G, Fig. 162, the rays pass through an adjustable slit, and are made parallel by the lens in the tube, B, before passing through the prism, P. The spectrum is seen through the telescope A. The tube, C, has at one end a scale on glass through which passes the light from a candle or coal-gas jet at F. This is reflected from the surface of the prism into the telescope A, where an image of the scale is seen alongside of the spectrum. Each part of the spectrum can thus be distinguished by its own scale number. Instead of a single prism, often a train of prisms is used, thus widening the spectrum and diminishing its brightness.*

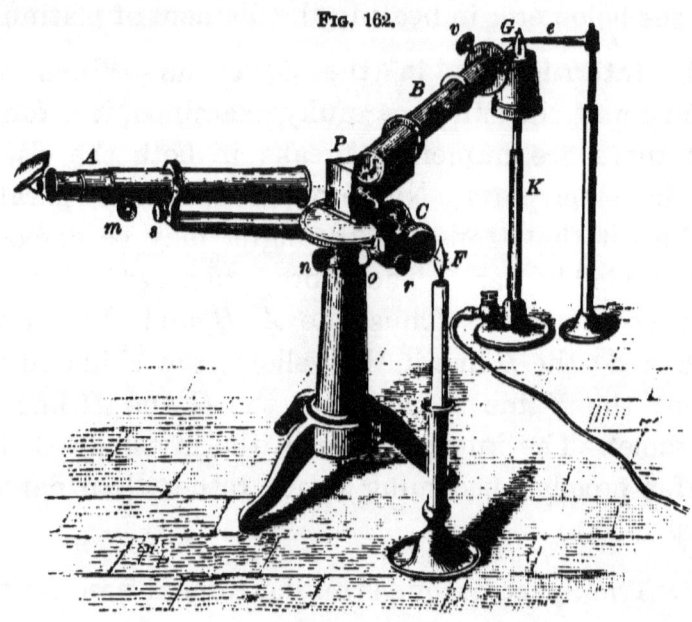

Fig. 162.

The Spectroscope.

6. Three Kinds of Spectra.—If in the spectroscope we examine the light of a glowing thin *gas or vapor*, its spectrum is seen to consist of one or more bright lines only. Thus burning sodium gives

---

* On the uses of the spectroscope, examine "New Astronomy," p. 258, and "Popular Chemistry," p. 147. In the former, opposite p. 258, is a colored illustration of the spectra.—The dark lines which cross the solar spectrum are known as *Fraunhofer's lines*, being so named in honor of the physicist who first carefully studied and mapped them. The spectroscope affords an unrivaled mode of analysis. No chemical test is so delicate. Strike together two books near the light at the slit of the spectroscope, and the dust blown into the flame will contain enough sodium (the basis of common salt) to cause the yellow D lines—its test—to flash out distinctly. A very effective spectroscope may be contrived thus: Cut a slit not over $\frac{1}{16}$ inch wide and 2 inches long in a piece of tin-foil, and gum it on a pane of glass. Hold this before a flame and look at it through a prism.

a pair of brilliant yellow lines close together; zinc vapor, a number of lines among which the blue are very prominent; and strontium, a number among which the red are conspicuous. Each element, if made gaseous, can be thus recognized by its spectrum.

If the light from a glowing solid be examined, its spectrum is found to be continuous, giving all the colors without interruption. White-hot lime, or the particles of carbon in a common candle-flame, furnish a continuous spectrum.

The bright lines given by a glowing gas may be made to broaden into bands, and these finally to become joined into a nearly continuous spectrum by subjecting the gas to very great pressure, and thus making it very dense. There is no sharp distinction between line spectra and continuous spectra.

The interrupted spectrum is that given by the sun and stars. It may be produced to a limited extent by interposing a glowing vapor, like that of sodium, between the spectroscope and a white-hot solid, like lime. It is believed that the body of the sun and of each of the stars is made up of very dense glowing matter, which is surrounded by less hot vapors. These absorb some of the light from within, and thus produce the interruptions observed in the spectra of the sun and stars. Moreover, it has been proved that each gas or vapor absorbs the same waves as those given out by itself in glowing. By comparing the bright lines of a known gas or vapor with the dark lines in the sun or star spectrum, it becomes possible to determine whether this vapor exists in the atmosphere of the sun or star.

7. Color.—If a piece of pure red paper is put against the successive parts of the spectrum on a screen, it will look red only when in the red part, but dark gray or black in the other parts. It reflects red light and absorbs the other tints. Color is analogous to pitch, violet corresponding to the high and red to the low sounds in music. Intensity of color, as of sound, depends on the amplitude of the vibrations. When a body absorbs all the colors of the spectrum except blue, but reflects that to the eye, we call it a blue body; when it absorbs all but green, we call it a green body.* Red glass has the power of absorbing all except the red rays, which it transmits. When a substance *reflects* all the colors to the eye, it seems to us white. If it *absorbs* all the colors, it is black. Thus color is not an inherent property of objects.† In darkness all things are colorless.

8. Complementary Colors.—Two colors, which by their mixture produce white light, are termed com-

---

* Some eyes are blind to certain colors, as some ears are deaf to certain sounds. "Color-blindness" generally exists as to red. Such a person can not by the color distinguish ripe cherries from green ones. Doubtless railway accidents have occurred through this inability to apprehend signals. Dr. Mitchell mentions a naval officer who chose a blue coat and red waistcoat, believing them of the same color; a tailor who mended a black silk waistcoat with a piece of crimson; and another who put a red collar on a blue coat. Dalton could see in the solar spectrum only two colors, blue and yellow, and having once dropped a piece of red sealing-wax in the grass, he could not distinguish it.

† Moisten a swab with alcohol saturated with common salt. On igniting this in a dark room, every object will take on a curious ghastly yellow hue from the burning sodium. The gay colors of flowers will instantly be quenched.

plementary to each other. Thus, if we sift the red rays out of a beam of light and bring the remainder to a focus, a bluish-green image will be formed.* In Fig. 163 the colors opposite each other are complementary. Place a red and a blue ribbon side by side. The former will take on a yellowish and the latter a greenish tint. Lay a piece of tissue paper upon black letters printed on brightly colored paper. The dark letters will appear of a color complementary to that of the background.†

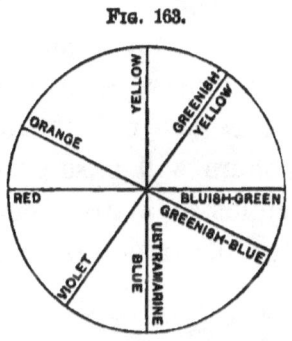

Fig. 163.

Complementary Colors.

9. The Rainbow is formed by the *refraction* and *reflection* of the sunbeam in drops of falling water. The white light is thus decomposed into its simple colors. The inner arch is termed the primary bow; the outer or fainter arch, the secondary.

PRIMARY Bow.—A ray of light, S'', enters, and is bent downward at the top of a falling drop, passes to the opposite side, is there reflected, then passing out of the lower side, is bent upward. By the refrac-

---

* Certain substances are able to split a ray of light into two colors, and are said to be *dichroic*. Gold-leaf reflects the yellow, transmits the green, and absorbs the rest.

† A color is heightened when placed near its complement. A red apple is the brighter for the contrast of the green leaf.—Observe a white cloud through a bit of red glass with one eye and through green glass with the other eye. After some moments, transfer both eyes to the red glass, opening and closing them alternately. The strengthening of the red color in the eye, fatigued by its complementary green, is very striking.—In examining ribbons of the same color, the eye becomes wearied and unable to detect the shade, because of the mingling of the complementary hue.

tion the ray of white light is decomposed, so that when it emerges it is spread out fan-like, as in the solar spectrum. Suppose that the eye of a spectator is in a proper position to receive the red ray, he can not receive any other color from the same drop, because the red is bent upward the least, and all the others will pass directly over his head. He sees the violet in a drop below. Intermediate drops furnish the other colors of the spectrum.

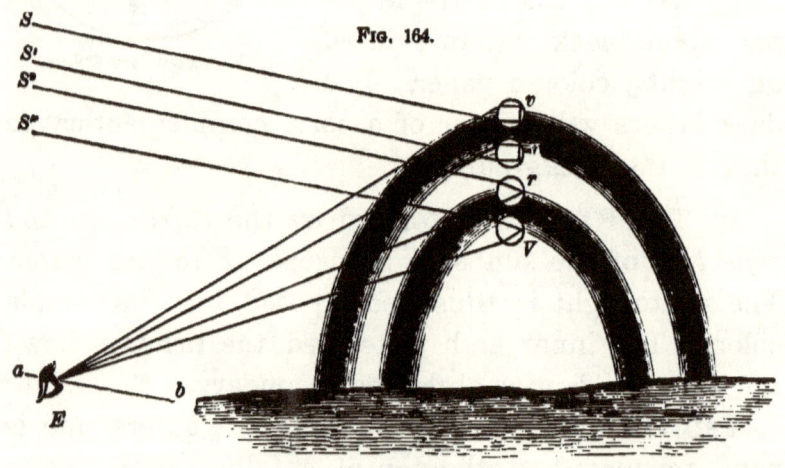

Fig. 164.

The Rainbow.

SECONDARY BOW.—A ray of light, S, strikes the bottom of a drop, v, is refracted upward, passes to the opposite side, where it is twice reflected, and thence passes out at the upper side of the drop. The violet ray being most refracted, is bent down to the eye of the spectator. Another drop, r, refracting another ray of light, is in the right position to send the red ray to the eye.

WHY THE BOW IS CIRCULAR.—When the red ray of

COMPOSITION OF LIGHT. 219

the primary bow leaves the drop, it forms an angle with the sun's ray, $S''r$, of about $42°$, and the violet forms with it one of $40°$. These angles are constant. Let ab be a straight line drawn from the sun through the observer's eye. If produced, it would pass through the center of the circle of which the rainbow is an arc. This line is termed the *visual axis*. It is parallel to the rays of the sun; and when it is also parallel to the horizon, the rainbow is a semicircle. Suppose the line EV in the primary bow to be revolved around Eb, keeping the angle bEV unchanged; the point V would describe an arc of a circle on the sky, and every drop over which it passed would be at the proper angle to send a violet ray to the eye at E. Imagine the same with the drop r. We can thus see (a) the bow must be circular; (b) when the sun is high in the heavens, the whole bow sinks below the horizon; (c) the lower the sun the larger is the visible circumference; and (d) on lofty mountains a perfect circle may sometimes be seen.*

10. Chromatic Aberration of Lenses.—Since in passing through any medium, such as a lens, the violet rays are bent farther away from their first direction than the red rays, they will be brought to a focus nearer than that of the red rays. An image on a screen produced with a single lens is therefore fringed with a reddish or a bluish fringe according

---

* Halos, coronas, sun-dogs, circles about the moon, and the tinting at sunrise and sunset, are produced by the refraction and reflection of the sun's rays by the clouds. The phenomenon known as the "sun's drawing water," consists of the long shadows of broken clouds. Twilight and kindred topics are treated in Astronomy.

to the position of the screen. By combining two lenses properly, one made of crown glass and the other of flint glass, it is possible to correct much of this coloring, which is called chromatic aberration, and at the same time much of the spherical aberration.*

11. Interference of Light (*Newton's Rings*).—Let the convex side of a plano-convex lens be pressed down upon a plane of glass. The two surfaces will apparently touch at the center. If different circles be described around this point, at all parts of each circle the surfaces will be the same distance apart, and the larger the circle the greater the distance. Now let a beam of red light fall upon the flat surface. A black spot is seen at the center; around this a circle of red light, then a dark ring, then another circle of red light, and so alternating to the circumference. The distances between the surfaces of the glass, where the successive dark rings appear, are proportional to the numbers 0, 2, 4 . . . . , and the bright circles to 1, 3, 5 . . . . . This fact suggests the cause. There are two sets of waves, one reflected from the upper surface of the plane glass, and the other from the lower surface of the

FIG. 165. 
Production of Newton's Rings.

---

* The crown-glass lens must be bi-convex and the flint-glass lens plano-concave or meniscus. Flint glass gives a spectrum nearly twice as long as crown glass. The two lenses oppose each other in their action on light. They may be so adjusted that each tends almost completely to reverse the spectrum that the other would produce, and yet the excess of deviation produced by the crown glass may still be enough to bring the rays of this nearly white light to a focus.

COMPOSITION OF LIGHT. 221

convex glass. These alternately interfere, producing darkness, and combine, making an intenser color.* To determine the length of a wave of red light, we have only to measure the distance between the two glasses at the first ring.

When beams of light of the various colors are used corresponding circles are obtained, having different diameters; red light gives the largest, and violet the smallest. We hence conclude that red waves are the longest, and violet the shortest. The minuteness of these waves passes comprehension. About 40,000 red waves, or 60,000 violet ones, are comprised within a single inch. Knowing the velocity of light, we can calculate how many of these tiny waves reach our eyes each second. When we look at a violet object, 757 million million of ether-waves break on the retina every moment!

12. Polarization of Light.—(1.) DEFINITION.—If we could look at the end of a ray of light coming to-

---

* The play of colors in mother-of-pearl is due to the interference of light in its thin overlapping plates.—In a similar manner the plumage of certain birds reflects changeable hues.—A metallic surface ruled with fine parallel lines not more than $\frac{1}{3000}$ of an inch apart, gleams with brilliant colors.—Thin cracks in plates of glass or quartz, mica when two layers are slightly separated, even the scum floating in stagnant water, breaks up the white light of the sunbeam and reflects the varying tints of the rainbow.—The rich coloring of a soap-bubble is caused by the interference of the rays reflected from the upper and lower surfaces of the bubble.—DIFFRACTION is interference produced by a beam of light passing along the edge of an opaque body or through a small opening, or reflected by a surface ruled with fine lines.—*Examples:* Place the blades of two knives closely together and hold them up to the sky; waving lines of interference will shade the open space.—Look at the sky through the meshes of a veil, or at a lamp-light through a bird-feather or a fine slit in a card, and delicate colors like those of the prism will appear.

ward us, as we can at the end of a rod, we should see the molecules of ether vibrating across the direction of the ray in all possible planes, as shown in Fig. 166. There are certain conditions under which reflected or refracted light may be made to vibrate in but a single plane. It is then called polarized light.

Fig. 166.

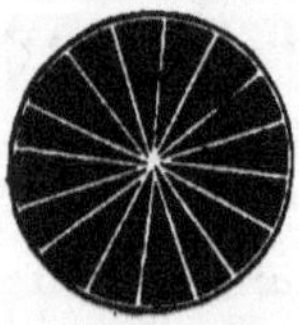

The crystal *tourmaline* has this power upon transmitted light. If two thin plates of this be cut parallel to the axis of the crystal and light be passed perpendicularly through them, when one is placed parallel to the other, as in Fig. 167, some of it will be absorbed, but what passes through vibrates only in a plane the same as that of the axis. This is proved by crossing them, as in Fig. 168; at once the light is quenched. What passed through the first plate had been polarized, and was stopped by the second plate when crossed. If they be placed with axes oblique to each other, part of the polarized light is transmitted and part quenched.

Fig. 167. Fig. 168. 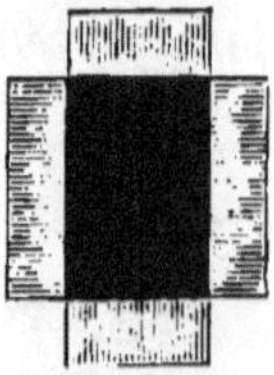
Tourmalines Parallel. Tourmalines Crossed.

(2.) DOUBLE REFRACTION.—In tourmaline and many other crystals the ether is unequally elastic in two directions at right angles to each other. The light is hence divided into two parts which pass through with unequal velocities. If transmitted across the axis of the crystal, these parts are separated so that two beams become perceptible. Iceland spar shows

this remarkably well. An object viewed through it appears double. If the crystal be placed over a dot and turned around, two dots will be seen; one appears a little nearer than the other and revolves around it, or a word will appear double if viewed in like manner. (Fig. 169.)
If now a plate of tourmaline be put between the eye and the rotating crystal of spar, the dots will alternately disappear. This shows that the two beams were polarized at right angles to each other.

Fig. 169.

Double Refraction.

One of them is called the *ordinary* and the other the extraordinary ray. Tourmaline is a doubly refracting crystal in which the ordinary ray is absorbed unless the plate be exceedingly thin.

(3.) POLARIZATION BY REFLECTION.—When light falls upon a surface of glass at such an angle that the reflected and refracted beams are at right angles to each other, each of these is polarized, just as in passing through a doubly-refracting crystal. This special polarizing angle of incidence for glass is about 56°. Many other substances polarize the light reflected at the proper angle from them.*

(4.) THE POLARISCOPE.—The best polarizer is a crystal of Iceland spar specially arranged so as to

---

* If a tourmaline is rotated before the eye while looking obliquely at the surface of a varnished table, or leather-seated chair, the reflected light will be found to be polarized.

transmit the extraordinary ray and quench the ordinary ray. It is called a Nicol's prism. Whatever is used for examining the light after it has been polarized is called an analyzer. The Nicol's prism makes the best analyzer also. An instrument that includes both polarizer and analyzer is called a polariscope. A glass plate fixed at the proper angle makes an excellent polarizer, and a small Nicol's prism, or piece of tourmaline, for analyzer is enough for many beautiful experiments. Exquisite displays of complementary colors, due to interference of polarized beams in transmission, may be seen by examining thin pieces of crystallized gypsum, mica, horn, strained glass, etc., between polarizer and analyzer.* Polarized light affords a delicate means of examining the molecular structure of a body.

* A simple polariscope is shown in Fig. 170. Upon a wooden frame a plate of glass, P, blackened on the under side, is fixed so that light falling on it at the polarizing angle shall be reflected through the tube tt'. This contains a small Nicol's prism, n, for analyzer, and a lens, l, through which an object, s, may be examined with polarized light. The student who

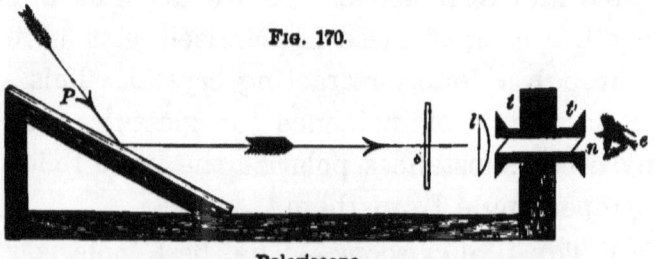

Fig. 170.

Polariscope.

makes one will find it a source of fascination and continued delight. He will find useful information very clearly expressed in the "Scientific American Supplement" for Nov. 20, 1886, p. 9072; also, a full and admirable explanation of these beautiful experiments in "Light," by Lewis Wright, published by Macmillan & Co., London.

V. OPTICAL INSTRUMENTS.

1. Microscopes (*to see small things*) are of two kinds, *simple* and *compound*. The former consists of one or more convex lenses through which the object

FIG. 171.

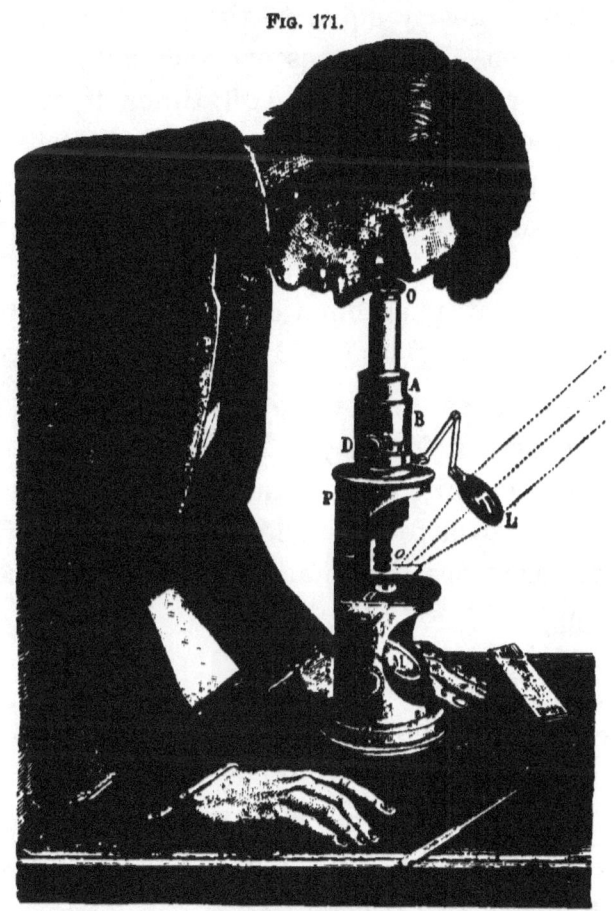

The Microscope.

is seen directly; the latter contains a simple magnifier for viewing the image of an object produced by a second lens. Fig. 171 represents a compound microscope. At M is a mirror which reflects the rays of light through the object a. The object-lens (objective), o, forms, in the tube above, a magnified, inverted image of the object. The eye-lens, O (ocular), magnifies this image. The magnifying power of the instrument is nearly equal to the product of that of the two lenses. If a microscope increases the apparent diameter of an object 100 times, it is said to have a power of 100 diameters, the surface being magnified $100^2 = 10,000$ times. The eye-piece may be only a single lens, and is really a simple microscope. The object-lens often consists of several lenses, and each one of a combination of convex crown glass and concave flint glass (p. 219) to prevent aberration.

2. Telescopes (*to see afar off*) are of two kinds, *reflecting* and *refracting*. The former contains a large metallic mirror (speculum) which reflects the rays of light to a focus. The observer stands at the side and examines the image with an eye-piece.*

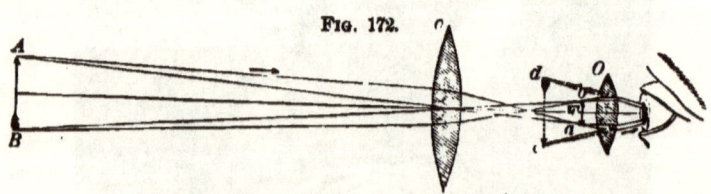

Fig. 172.

Formation of Image in the Telescope.

The Refracting Telescope contains an object-lens, o, which forms an inverted image, ab. This is viewed through the eye-piece, O, which produces a magnified

---

* The largest reflecting telescope is that of Lord Rosse (see Frontispiece to "Astronomy"). Its speculum is 6 ft. in diameter and gathers about 120,000 times as much light as would ordinarily enter the eye.

image, *cd*, of the first image, *ab*. The image *cd* is as much larger than *ab* as the focal distance of the object-glass exceeds that of the eye-glass. The larger

Fig. 173.

Cambridge Equatorial.

the object-lens the more light is collected with which to view the image. The magnifying power is due to the eye-piece.* The apparent inversion of the ob-

---

* The use of the telescope depends upon (1st) its light-collecting and (2d) its magnifying power. Thus Herschel, illustrating the former point, says that once he told the time of night from a clock on a steeple invisible on account of the darkness. It is noticeable that while in the compound microscope the image is as much larger than the object as the image is farther than the object from the object-glass, in the telescope the image is as much smaller than the object as it is nearer than the object to the object-glass; while in both cases the image is examined with a magnifier. If

ject is of no importance for astronomical purposes. In terrestrial observations additional lenses are used to erect the image.

3. The Opera-glass contains an object-glass, O, and an eye-piece, o. The latter is a double-concave lens; this increases the visual angle by diverging

Fig. 174.

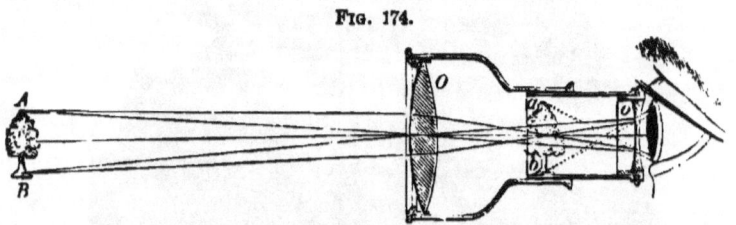

Formation of Image in Opera glass.

the rays of light, which would otherwise come to a focus beyond the eye-piece. An erect and magnified image is seen at ab.

4. The Projecting Lantern consists of a system of lenses attached to a dark box, within which is a powerful source of illumination such as the electric light or lime light. Sometimes an oil-lamp is used. From the white-hot source the light is converged by the condensing lens, C, Fig. 175, so as to send through the projecting lens, P, as much of it as possible. A picture on glass is placed in front of the

---

a power of 1,000 be used in looking at the sun, we shall evidently see the sun as if it were only 93,000 miles away, or less than one half the distance of the moon. The same power used upon the moon would bring that body apparently to within 240 miles of us.

The National Observatory telescope at Washington has an object-glass 26 inches in diameter, and of excellent defining power. The Lick telescope, erected in 1887 upon Mt. Hamilton, in California, has an object-glass just one yard in diameter. Its light-collecting power is estimated to be about 30,000 times that of the unaided eye.

condenser, and is thus strongly illuminated. An image of it, greatly enlarged, is formed by the pro-

Fig. 175.

Projecting Lantern for the Lime Light.

jecting lens and focalized on a distant white screen in a dark room. *Dissolving views* are produced by

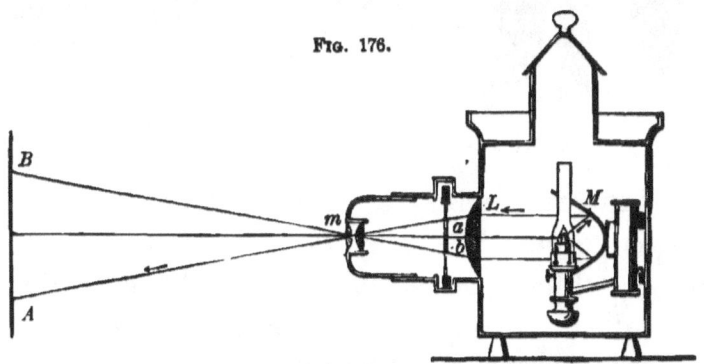

Fig. 176.

Projecting Lantern for the Oil Light.

using two lanterns together. While one view is on the screen, another is projected upon it. The light

230　　　　　　　　OPTICS.

is then cut off from the first lantern so as to leave only the second view. Fig. 176 is an outline of one form of oil-lantern. The reflector, M, helps to illuminate the transparent picture, ab, in front of the condenser.

5. The Camera, used by photographers, contains a double-convex lens, L, which throws an inverted image of the object upon a removable groundglass screen, S. When the focus has been obtained, the screen is removed and a slide, containing a sensitive film, is inserted in its place. ("Chemistry," p. 171.)

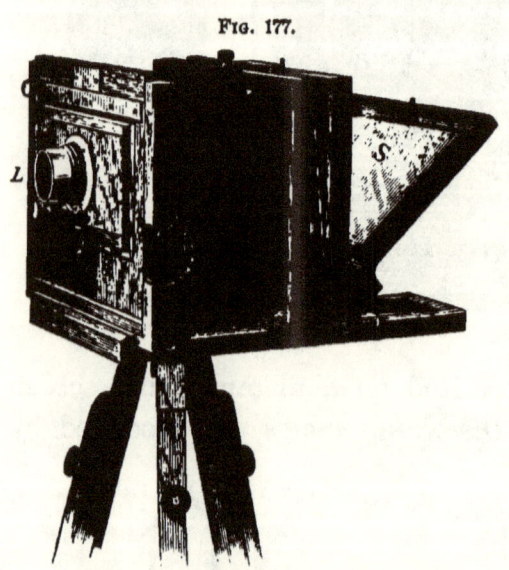

Fig. 177.

Photographer's Camera.

6. The Eye is a unique optical instrument resembling a camera. The outer membrane is termed the *sclerotic* coat, S (Fig. 178). It is tough, white, opaque, and firm. A little portion in front, called the *cornea*, c, is more convex and perfectly transparent. The middle or *choroid* coat, C, is soft and delicate, like velvet. It lines the inner part of the eye and is covered with a black pigment. Over it the optic nerve, which enters at the rear, expands

OPTICAL INSTRUMENTS. 231

in a net-work of delicate fibers termed the *retina*, the seat of vision. Back of the cornea is a colored curtain, *hi*, the *iris* (rainbow), in which is a round hole called the *pupil*. The *crystalline lens*, *o*, is a double-convex lens, composed of concentric layers somewhat like an onion, weighing about four grains and transparent as glass. Between the cornea and the crystalline lens is a limpid fluid termed the *aqueous humor;* while the *vitreous humor*, a transparent, jelly-like liquid, fills the space back of the crystalline lens.

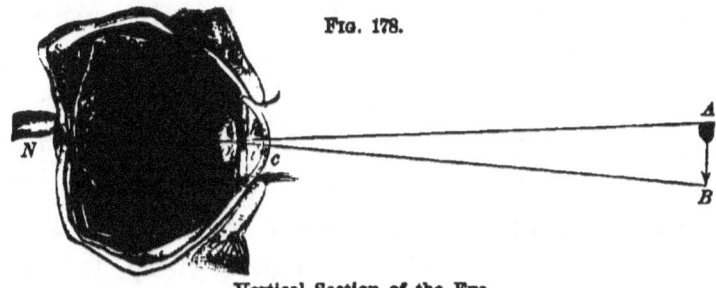

Fig. 178.

Vertical Section of the Eye.

Let *AB* represent an object in front of the eye. Rays of light are first refracted by the cornea and aqueous humor, next by the crystalline lens, and last by the vitreous humor, forming on the retina an image, *ab*,* which is real, inverted, and smaller than the object. To render vision distinct, the rays must

---

* The diameter of the eye is less than an inch; yet, as we look over an extended landscape, every feature, with all its variety of shade and color, is repeated in miniature on the retina. Millions upon millions of ether waves, converging from every direction, break on that tiny beach, while we, oblivious to the marvelous nature of the act, think only of the beauty of the revelation. Yet in it the physicist sees a new illustration of the simplicity and perfection of the laws and methods of the Divine Workman, and a continued reminder of His forethought and skill.

be accurately focused on the retina. If we gaze steadily at an object near by, and at the same time regard a distant object in the same direction, we find our vision of this blurred. If now we gaze at the more distant object, our vision of the nearer one becomes blurred. The eye thus has the power of adapting itself to the varying distances of objects. This is done by a change in the convexity of the front surface of the crystalline lens under the action of the ciliary muscle which surrounds it at its edge. When clear vision can not be had of distant objects, the person is *near-sighted*. When the ciliary muscle is strained to produce clear vision of objects less than ten or twelve inches distant, the person, if young, is *over-sighted*. In the first case, the distance of the retina from the crystalline lens is too great to permit of distinct focalization; in the second case, this distance is too small. The remedy for near-sightedness is to wear concave glasses, selected by a competent oculist. Rays from distant objects are thus made to diverge before entering the eye, as if they had come from very near objects. For over-sightedness, the remedy is properly selected convex glasses.* As old age approaches, the crystalline lens

---

* There are other defects for which the aid of the oculist should be sought. If glasses are needed, they should never be selected except after examination by a thoroughly competent person. If not properly adapted, they may do much more harm than good. No eye is optically perfect, and but few are free from defects that may be detected on examination. That glasses are more used now than during the previous generations is due not so much to increase of habits injurious to vision as to the better knowledge of the eye and the better opportunities for every person to find out his own defects.

becomes less elastic so that the eye loses the power of accommodation to near objects. Convex glasses become necessary for reading, while the vision of distant objects may remain perfect.

The retina retains an impression for a brief time after the object has been removed, usually a fraction of a second, which varies according to the brightness.* This explains why a lighted coal, rapidly moved in the dark, appears as a line of light. Many of the most brilliant effects from fire-works depend on this property of the retina. The *Zoetrope* is an instrument by which a succession of pictures of the same object in different phases of motion are made to pass rapidly before the eye. The persistence of the successive sensations causes an apparent blending, so that the illusion is that of an object actually in motion.

7. Binocular Vision.—In looking with both eyes at an object that is not very distant, we obtain a much better idea of its position and form, or its "depth in space," than when a single eye is employed. The two retinal images differ slightly because the two eyes are different in direction from the object, and hence to a slight extent we see around the object on two sides. The illusion of depth in space is well brought out by means of Fig. 179. The tunnel, *A*, appears as if viewed by the left eye alone; *B*, as if

---

* When one is riding slowly on the cars and looking at the landscape between the upright fence-boards, he catches only glimpses of the view; but *when moving rapidly, these snatches will combine to form a perfect landscape*, which has, however, a grayish tint, owing to the decreased amount of light reflected to the eye.

by the right eye alone. Bring the page close up to the face, so that one picture is immediately in front of each eye. The two images at once seem combined into a single blurred image. Now withdraw the page a few inches;* the haziness gives place to distinct-

Fig. 179.

A Stereograph which may be Viewed without a Stereoscope.

ness and the tunnel appears startlingly deep, as if it were a hole through the book. While still gazing into its depths two more tunnels may be indirectly seen, one on each side of it; but the illusion of depth in them is far less clear. Each is seen by but a single eye, while the middle one is a binocular perception.

The *Stereoscope* is an instrument intended to aid in attaining binocular vision of a pair of properly prepared pictures, which together compose the stereograph. With a little practice like that just described, any one may become independent of the stereoscope.

---

* In performing this experiment, it is very important to avoid crossing the eyes. Perfect relaxation of the muscles of the eyeballs will make it very easy. Imagine yourself to be looking *through* the page at the opening of a *distant* tunnel, and keep the muscles relaxed.

For further discussion of the Stereoscope, consult an article on this subject in the "Popular Science Monthly" for May and June, 1882.

PRACTICAL QUESTIONS.

1. Why is the secondary bow fainter than the primary? Why are the colors reversed?
2. Why can we not see around the corner of a house, or through a bent tube?
3. What color would a painter use if he wished to represent an opening into a dark cellar?
4. Is white a color? Is black?
5. By holding an object nearer a light, will it increase or diminish the size of the shadow?
6. What must be the size of a glass in order to reflect a full-length image of a person? *Ans.* Half the person's height.
7. Where should we look for a rainbow in the morning?
8. Can two spectators see the same bow?
9. Why, when the drops of water are falling through the air, does the rainbow appear stationary?
10. Why can a cat see in the night better than a human being?
11. Why can not an owl see distinctly in daylight?
12. Why are we blinded when we pass quickly from a dark into a lighted room?
13. If the light of the sun upon a distant planet is $\frac{1}{100}$ of that which we receive, how does its distance from the sun compare with ours?
14. If, when I sit six feet from a candle, I receive a certain amount of light, how much shall I diminish it if I move back six feet farther?
15. Why do drops of rain, in falling, appear like liquid threads?
16. Why does a towel turn darker when wet?
17. Does color exist in the object, or in the mind of the observer?
18. Why is lather opaque, while air and a solution of soap are each transparent?
19. Why does it whiten molasses candy to "pull it"?
20. Why does plastering become lighter in color as it dries?
21. Why does the photographer use a lamp with a chimney of red glass in the "dark room"?
22. Is the common division of colors into "cold" and "warm" verified in philosophy?
23. Why is the image on the camera, Fig. 177, inverted?
24. Why is the second image seen in a mirror, Fig. 140, brighter than the first?
25. Why does a blow on the head make one "see stars"? *Ans.* The blow excites the optic nerve, and so produces the sensation of light.
26. What is the principle of the kaleidoscope? (If you can not discover this, consult Deschanel's "Natural Philosophy," pp. 886–891.)
27. Which can be seen at the greater distance—gray or yellow?
28. When a star is near the horizon, does it seem higher or lower than its true place?

29. Why can we not see a rainbow at midday?

30. What conclusion do we draw from the fact that moonlight shows the same dark lines in the spectrum as sunlight?

31. Why does the bottom of a boat seen under clear water appear flatter than it really is?

32. Of what shape does a round body appear in water?

33. Why is rough glass translucent while smooth glass is transparent?

34. Why can a carpenter, by looking along the edge of a board, tell whether it is straight?

35. Why can we not see out of the window after we have lighted the lamp in the evening?

36. Why does a ground-glass globe soften the light?

37. Why can we not see through ground-glass or painted windows?

38. Why does the moon's surface appear flat?

39. Why can we see farther with a telescope than with the naked eye?

40. Why is not snow transparent, like ice?

41. Are there rays in the sunbeam which we can not perceive with the eye?

42. Why, when we press the finger on one eyeball, do we see objects double?

43. Why does a distant light, in the night, seem like a star?

44. Why does a bright light, in the night, seem so much nearer than it is?

45. What color predominates in artificial lights? *Ans.* Yellow.

46. Why are we not sensible of darkness when we wink?

47. Under what condition do the eyes of a portrait seem to follow a spectator to all parts of a room?

48. Why do the two parallel tracks of a railroad appear to approach in the distance?

49. Why does a fog apparently magnify objects?

50. If you sit where you can not see another person's image, why can not that person see yours?

51. Why can we see the multiple images in a mirror better if we look into it very obliquely?

52. Why is an image seen in water inverted?

53. Why is the sun's light fainter at sunset than at midday?

54. Why can we not see the fence-posts when we are riding rapidly?

55. Ought a red flower to be placed in a bouquet close to an orange one? A pink or blue with a violet one?

56. Why are the clouds white while the clear sky is blue?

57. Why does skim-milk look blue and new milk white?

58. Why is not the image of the sun in water at midday so bright as near sunset?

59. Why is the rainbow always opposite the sun?

60. Hold a card with its edge close in front of your eye and look at a distant candle flame in a dark room. You will probably perceive either a reddish or a bluish fringe on one side. Explain.

SUMMARY.

Light comes from the sun and other self-luminous bodies. It is transmitted by means of vibrations in ether, in accordance with the laws of wave-motion. It is radiated equally in all directions, travels in straight lines, decreases as the square of the distance increases, and is propagated 186,000 miles per second. Light falling upon a body may be absorbed, transmitted, or reflected. If the surface be rough, the irregularly-reflected light enables us to see the body; if it be smooth and highly polished, the rays are reflected so as to form an image of the original object. Surfaces producing such images are termed mirrors—plane, concave, or convex. The image is seen in the direction from which the reflected ray enters the eye, and, in a plane mirror, as far behind the mirror as the object is in front. Multiple images are produced by repeated reflections, as in the kaleidoscope. A concave mirror, as generally used, collects the rays, and serves to produce either a magnified erect virtual image or a magnified or diminished inverted real image of an object. A convex mirror scatters the rays, and diminishes the apparent size of an object.

When a ray enters or leaves a transparent body obliquely, it is refracted; if passing into a rarer medium, it is bent away from the perpendicular erected at the point of incidence; if into a denser medium, it is bent toward this perpendicular. A lens is a transparent body with one or more curved surfaces, which are usually spherical, so as to refract the light either to a focus, or as if it had come from a focus. There are two classes—convex and concave. The former lens, as generally used, tends, like a concave mirror, to collect the rays of light; the latter, like a convex mirror, causes the rays of light to diverge. Mirage is an optical delusion caused by refraction of light in passing through air composed of strata of unequal density. Owing to the varying refrangibility of the different waves of the sunbeam, a prism can disperse them into a colored band called the solar spectrum. The spectrum shows white light to consist of many tints, and that the solar energy may produce luminous, heating, or chemical effects according to the nature of the .

body receiving it. By means of the spectroscope we can examine the spectrum of a flame, and find whether its light is due to the incandescence of a gas or to the glowing of solid particles disseminated through it. Each substance in the gaseous state gives a spectrum with its peculiar lines of color. A gas absorbs the same rays that it is capable of emitting; if, therefore, a burning gas or vapor is interposed between the eye and a glowing solid, the spectrum of the solid is interrupted by dark lines due to absorption by the vapor. A delicate mode of analysis is thus furnished, whereby the elements even of the distant stars can be detected. The rainbow is formed by the refraction and reflection of the sunbeam in rain-drops. Light, when reflected by or transmitted through bodies, is so modified, chiefly by absorption, as to produce the varied phenomena of color. Each color has its own wave-length, which is less than $\frac{1}{35000}$ inch. Different systems of light-waves, as of sound-waves, may be combined. But if any two coincide with similar phases they will strengthen each other; and if with opposite phases, weaken each other. Interference of light, as thus produced, causes the play of colors in the soap-bubble, mother-of-pearl, etc. Polarized light is that in which the molecular vibrations are made in the same plane. Many of the most beautiful color effects may be produced by polarization.

The principal optical instruments, including the eye, are adapted to produce and examine the image formed by a lens. In the projecting lantern and solar microscope, the image is thrown on a screen in a dark room. In the refracting telescope and the microscope, the image is formed in a tube by a lens at one end and looked at from behind by a lens at the other end. In the ey , which is a small camera-obscura, the image is formed on the retina, whence the sensation is carried by the optic nerve to the brain. The retinal sensation continues for a short time after the impression is made. Advantage is taken of this fact in the use of the zoetrope, by which a succession of images is made to appear in motion. Vision with two eyes is superior to that with a single eye, because we are thus enabled to form better ideas of depth in space, and hence of the distance and form of a body. The stereoscope is an instrument for studying the peculiarities of binocular vision.

HISTORICAL SKETCH.

THE ancients knew that light is propagated in straight lines. They discovered the laws of reflection, and one of the ancient fables is that Archimedes set fire to the Roman ships off Syracuse by means of concave mirrors. Euclid and Plato, however, thought that the ray of light proceeds from the eye to the object, an error that was long uncorrected. One thousand years did not bring much advancement in this department of knowledge. The Arabian philosopher, Alhazen, who lived in the eleventh century, discovered the apparent displacement of a body seen in water. The law of intensity of light was established by Kepler, and the first researches on the comparison of intensity from different sources were made by Maurolycus, Huygens, and Francis Marie. About 1608, the telescope was invented by the Dutch.* Jansen, Metius, and Lippersheim each claimed the honor, and the legend is that the discovery grew out of some children at play, accidentally arranging two watch-glasses so as apparently to magnify an object. In fact, however, the action of the convex lens was already known, the compound microscope had been invented by Jansen twenty years previously, and the simple microscope was known to the ancient Chaldeans. In 1621, Snell discovered the law of refraction. By its aid Descartes explained the rainbow. Half a century of waiting, and Newton published his investigations in the decomposition of light. He, however, believed in what is known as the "corpuscular theory." This holds that light consists of minute particles of matter radiated in straight lines from a luminous object, the ray being endued with alternate "fits" of easy reflection and easy transmission. In 1676, Roemer, by observing Jupiter's moons (p. 192), found out the velocity of light, which up to that time had been considered instantaneous. In 1665, Gri-

---

* "In 1609, the government of Venice made a considerable present to Signor Galileo, of Florence, Professor of Mathematics at Padua, and increased his annual stipend by 100 crowns, because, with diligent study, he found out a rule and measure by which it is possible to see places 30 miles distant as if they were near, and, on the other hand, near objects to appear much larger than they are before our eyes."—*From an old paper in the Library of Heidelberg University.*

maldi discovered the existence of fringes of light and shade when a beam is received through a narrow slit. Huygens soon afterward advanced the undulatory theory, which was originated independently about the same time by Hooke. This involved them in vigorous disputes with Newton, without the definite establishment of their theory. In 1802, Thomas Young revived the undulatory theory, accounting by it for all the phenomena of interference then known. In 1817, Fresnel extended the researches of Young, and Newton's corpuscular theory began to fall into discredit. The elementary phenomena of polarization were discovered by Malus in 1808, and this subject was afterward studied with great thoroughness by Fresnel, Arago, Biot, and Brewster.

VIII.

ON HEAT.

"THE combustion of a single pound of coal, supposing it to take place in a minute, is equivalent to the work of three hundred horses; and the force set free in the burning of 300 lbs. of coal is equivalent to the work of an able-bodied man for a life-time."

ANALYSIS OF HEAT.

HEAT.

- **I. Production of Heat.**
  1. Definitions.
  2. Relation between the Forms of Radiant Energy.
  3. Theory of Heat.
  4. Sources of Heat.
  5. Mechanical Equivalent of Heat.

- **II. Physical Effects of Heat.**
  1. Expansion.
  2. Temperature.
  3. The Heat Unit.
  4. Liquefaction.
  5. Vaporization.
  6. Evaporation.
  7. Spheroidal State.
  8. Specific Heat.

- **III. Communication of Heat.**
  1. Conduction.
  2. Convection.
  3. Radiation.
  4. Absorption and Reflection.

- **IV. The Steam-Engine.**
  1. General Principle.
  2. The Governor.
  3. The High-pressure Engine.

- **V. Meteorology.**
  1. General Principles.
  2. Dew.
  3. Fogs.
  4. Clouds.
  5. Rain.
  6. Winds.
  7. Ocean Currents.
  8. Adaptations of Water.

HEAT.

I. PRODUCTION OF HEAT.

1. Definitions.—*Radiant Energy* is the name of what we receive from the sun, stars, and other heated bodies. It may be manifested as light, as temperature, as chemism, or in all of these ways at the same time.

2. Relation between the Forms of Radiant Energy.—Thrust a cold iron into the fire. It is at first dark, but soon becomes luminous, like the glowing coals.—Raise the temperature of a platinum wire. We quickly feel the radiation of obscure heat-rays. As the metal begins to glow, our eyes detect a red color, then orange combined with it, and so on through the spectrum. At last all the colors are emitted, and the metal is dazzling white. Like light, heat may be reflected, refracted, and polarized. It radiates in straight lines in every direction, and decreases in intensity as the square of the distance increases. It moves with the same velocity as light.

It is believed that each of the forms of radiant energy is merely the manifestation of wave-motion at a special rate.* The longer and slower waves of

* According to Tyndall, 95 per cent. of the rays from a candle are

ether falling upon the nerves of touch produce the sensation of heat. The more rapid affect the optic nerve and produce the sensation of light. The shortest are especially active in producing chemical changes.

3. Theory of Heat.—Heat is motion. The mole-

invisible or heat-rays. These may be brought to a focus and bodies fired in the darkness.—Each of the five classes of nerves seems to be adapted to transmit vibrations of its own kind, while it is insensible to the others. Thus, if the rate of oscillation be less than that of red, or more than that of violet, the optic nerve is uninfluenced by the waves. We can not see with our fingers, taste with our ears, or hear with our nose. Yet these are organs of sensation and sensitive to their peculiar impressions.—" Suppose, by a wild stretch of imagination, some mechanism that will make a rod turn round one of its ends, quite slowly at first, but then faster and faster, till it will revolve any number of times in a second; which is, of course, perfectly imaginable, though you could not find such a rod or put together such a mechanism. Let the whirling go on in a dark room, and suppose a man there knowing nothing of the rod; how will he be affected by it? So long as it turns but a few times in the second, he will not be affected at all unless he is near enough to receive a blow on the skin. But as soon as it begins to spin from sixteen to twenty times a second, a deep growling note will break in upon him through his ear; and as the rate then grows swifter, the tone will go on becoming less and less grave, and soon more and more acute, till it will reach a pitch of shrillness hardly to be borne, when the speed has to be counted by tens of thousands. At length, about the stage of forty thousand revolutions a second, more or less, the shrillness will pass into stillness; silence will again reign as at first, nor any more be broken. The rod might now plunge on in mad fury for a long time without making any difference to the man; but let it suddenly come to whirl some million times a second, and then through intervening space faint rays of heat will begin to steal toward him, setting up a feeling of warmth in his skin; which again will grow more and more intense, as now through tens and hundreds and thousands of millions the rate of revolution is supposed to rise. Why not billions? The heat at first will be only so much the greater. But, lo! about the stage of four hundred billions there is more—a dim red light becomes visible in the gloom; and now, while the rate still mounts up, the heat in its turn dies away, till it vanishes as the sound vanished; but the red light will have passed for the eye into a yellow, a green, a blue, and, last of all, a violet. And to the violet, the revolutions being now about eight hundred billions a second, there will succeed darkness—night, as in the beginning. This darkness too, like the stillness, will never more be broken. Let the rod whirl on as it may, its doings can not come within the ken of that man's senses."

cules of a solid are in constant vibration. When we *increase* the rapidity of this oscillation, we heat the body; when we *decrease* it, we cool the body. The vacant spaces between the molecules are filled with ether. As the air moving among the limbs of a tree sets its boughs in motion, and in turn may be kept in motion by the waving of branches, so the ether puts the molecules in vibration, or is thrown into vibration by them.—*Example:* Insert one end of a poker in the fire. The particles immersed in the flame are made to vibrate intensely; the swinging molecules strike their neighbors, and so on, continually, until the oscillation reaches the other end. If we handle the poker, the motion is imparted to the delicate nerves of touch; they carry it to the brain, and pain is felt. In popular language, "the iron is hot," and we are burned. If, without touching it, we hold our hand near the poker, the ether-waves set in motion by the vibrating molecules of iron strike against the hand, and produce a less intense sensation of heat. In the former case, the fierce motion is imparted directly; in the latter, the ether acts as a carrier to bring it to us.

4. The Sources of Heat are the sun, the stars, and mechanical and chemical energy.

(1.) The molecules of the sun and stars are in rapid vibration. These set in motion waves of ether, which are propagated across the intervening space, and meeting the earth, give up their motion to it. (2.) Friction and percussion produce heat, the motion of a mass being changed into motion among molecules.* (3.) Chemical action is seen in fire. The oxygen of the air has an affinity for the carbon and hydrogen of the fuel. They combine, and chemical energy is transformed into that of sensible heat.

5. Mechanical Equivalent of Heat (*Joule's Law*).—In these various changes of mechanical motion into motion of molecules no energy is destroyed, though some of it may be so transformed as to become incapable of being made to do *useful* work. If the energy transformed by the fall of a blacksmith's hammer on his anvil could be gathered up, it would be sufficient to lift the hammer to the point from which it fell. *A pound-weight falling vertically 772 feet, will generate enough heat to raise the temperature of 1 pound of water through 1° F.;* conversely, this amount of heat is the equivalent of the energy required to lift 1 pound mechanically to a height of 772 feet. This important truth was first demonstrated by Mr. Joule, of Manchester, England, and we express it by saying 772 foot-pounds is the mechanical equivalent of heat. Expressed in metric measures, it is 424 kilogram-meters for 1° C.

---

* A horse hits his shoes against a stone and "strikes fire"; little particles of the metal being torn off are heated by the shock, and some of the energy is manifested also as light.—A train of cars is stopped by the pressure of the brakes. In a dark night, we see the sparks flying from the wheels, the motion of the train being converted into heat.—A blacksmith pounds a piece of iron until it glows. His strokes set the particles of metal vibrating rapidly enough to send ether-waves of such swiftness as to affect the eye of the observer.—As a cannon-shot strikes an iron target, a shower of sparks is scattered around.—Were the earth instantly stopped, enough heat would be produced to "raise a lead ball the size of our globe to 384,000° C." If it were to fall to the sun its impact would produce a thousand times more heat than its burning.

II. PHYSICAL EFFECTS OF HEAT.

1. Expansion.—If the molecules of a body have an increase of energy imparted to them they swing, like pendulums, through wider arcs. Each tends to push against its neighbor, and the mass as a whole grows larger. Hence the general law, "Heat expands and cold contracts," cold being merely a relative term implying the withdrawal of energy. The ratio of the increase of volume to the original volume for a change of 1° in temperature is called the *Co-efficient of Expansion*. Generally this is greatest for gases, less for liquids, and least for solids, each particular substance having its own co-efficient. The force of expansion is for many substances irresistible. A rise in temperature of 80° F. will lengthen a bar of wrought-iron, 10 feet long, about $\frac{1}{17}$ of an inch; and if its cross-section is one square inch it will push in expanding with a force of about 25 tons. When the metal cools it will contract with the same force.*

A familiar application of expansion is in the pen-

---

* A carriage-tire is put on when hot, in order that, when cooled, it may bind the wheel together.—Rivets used in fastening the plates of steam-boilers are inserted red-hot.—"The ponderous iron tubes of the Britannia Bridge writhe and twist, like a huge serpent, under the varying influence of the solar heat. A span of the tube is depressed only a quarter of an inch by the heaviest train of cars, while the sun lifts it 2½ inches." The same may be noticed on the great Brooklyn Bridge, more than a mile long, where an allowance of nearly a yard has to be made for expansion with the change of seasons.—The Bunker-hill monument nods as it follows the sun in its daily course.—Tumblers of thick glass break on the sudden application of heat, because the surface dilates before the heat has time to be conducted to the interior.

248   HEAT.

dulum of a clock, which lengthens in summer and shortens in winter. A clock, therefore, tends to lose time in summer and gain in winter. To regulate it we raise or lower the pendulum bob.

Fig. 180.

The *gridiron pendulum* consists of brass and steel rods, so connected that the brass, h, k, will lengthen upward, and the steel, a, b, c, d, downward, and thus the center of oscillation remain unchanged. The *mercurial pendulum* contains a cup of mercury which expands upward, while the pendulum-rod expands downward.

2. Temperature.—When one body is in a condition to communicate heat to another, the first is said to have a higher *temperature* than the second, or to be warmer. We measure temperature usually by noting its effect in producing expansion. Within narrow limits we may form a rough estimate of it by the sensation of touch, but this is very unreliable.

Gridiron Pendulum.

The thermometer is an instrument for measuring temperature, usually by the expansion of mercury.*

---

* Take a glass tube terminating in a bulb, and heat the bulb to expel the air. Then plunge the stem in colored water. As the bulb cools, the water will rise and partly fill it. Heat the bulb again until the steam pours out of the stem. On inserting it a second time, the water will fill the bulb. In the manufacture of thermometers, it is customary to have a cup blown at the upper end of the stem. This is filled with mercury, and

To graduate it, according to Fahrenheit's scale (F.), each thermometer is put in melting ice, and the point to which the mercury sinks is marked 32°, *Freezing-point.** It is then placed in a steam-bath, and the point to which the mercury rises (when the barometric column stands at 30 inches) is marked 212°, *Boiling-point.* The space between these two points is divided into 180 equal parts. In the Centigrade scale (C.) the freezing-point is 0, and the boiling-point 100°. In Reaumur's scale (R.), the boiling-point is 80°.† The thermometer does not measure the quantity of heat, but only its intensity.

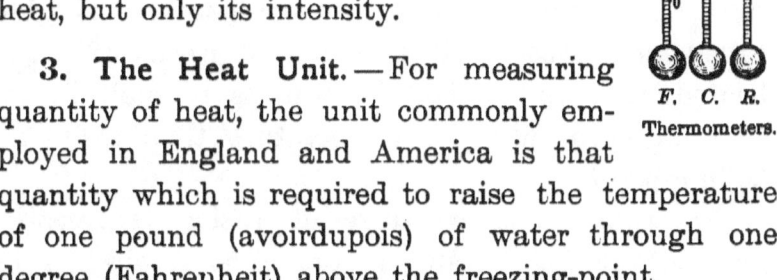

Fig. 181.

F. C. R.
Thermometers.

3. The Heat Unit.—For measuring quantity of heat, the unit commonly employed in England and America is that quantity which is required to raise the temperature of one pound (avoirdupois) of water through one degree (Fahrenheit) above the freezing-point.

---

the air, when expanded, bubbles out through it, while the metal trickles down as the bulb cools. The mercury is then highly heated, when the tube is melted off and sealed at the end of the column of mercury. The metal contracts on cooling, and leaves a vacuum above.

* The inventor placed zero 32° below the temperature of freezing water, because he thought that to be absolute cold—a point now estimated to be about 492° below the freezing-point on his scale.

† The following formulæ will be of use in comparing the readings of the different scales:

$$R. = \tfrac{4}{9}\, C. = \tfrac{4}{9}\, (F. - 32°) \quad \ldots \ldots \ldots (1.)$$
$$C. = \tfrac{5}{4}\, R. = \tfrac{5}{9}\, (F. - 32°) \quad \ldots \ldots \ldots (2.)$$
$$F. = \tfrac{9}{5}\, C. + 32° = \tfrac{9}{4}\, R. + 32° \quad \ldots \ldots \ldots (3.)$$
$$1° \; C. = 1.8° \; F. \quad \ldots \ldots \ldots \ldots \ldots (4.)$$

4. Liquefaction or Fusion.—When heat is communicated to a solid body a point is finally reached when the vibratory swing of its molecules is so great that they are driven apart, each toward the limit of the sphere of attraction of its neighbor, so that all rigidity is lost.* The molecules then move freely among themselves. The energy that is applied raises the temperature of the body up to a fixed point called its melting or fusing point, when liquefaction begins. Additional energy then does the work of driving the molecules apart without further rise of temperature, until fusion is complete; after which the liquid rises still further in temperature. Energy that does thus the work of changing the state of a body without at the same time changing its temperature is often called *latent heat*.† If a pound of ice at 32° F. be heated, it requires 142 heat units to melt it, and 180 more to raise its temperature then up to the boiling-point.

Freezing.—The converse of fusion is freezing. Ice melts at 32° F., and in doing so it absorbs energy. Water freezes at 32° F., and in doing so it gives out the energy which had been keeping its molecules apart. Thawing is thus a cooling process and freezing is a warming process. Freezing mixtures depend on this principle. In freezing ice-cream, salt and pounded ice are put around the

---

* This is true only of bodies which are not broken into their chemical constituents before the melting-point is reached. A large variety of substances, such as wood, bone, flesh, etc., become chemically changed instead of melting.

† The term *latent heat* is gradually going out of use.

PHYSICAL EFFECTS OF HEAT. 251

vessel that contains the cream. The strong attraction between salt and water causes the ice to melt rapidly, and the solid salt becomes liquid by solution. This rapid thawing involves much absorption of energy, which comes from the nearest objects whose temperature is higher than that of the solution. The cream thus loses energy, its temperature becoming reduced down to the freezing-point.*

5. Vaporization. — When heat is applied to a liquid the temperature rises until the boiling-point is reached, when it stops and the liquid is changed to vapor at that constant temperature. The vapor is nearly free from solids dissolved in the liquid.— *Example:* Pure or distilled water is obtained by heating water in a boiler, A, whence the steam passes through the pipe, C, and the *worm* within the condenser, S, where it is condensed and drops into the vessel, D. The pipe is coiled in a spiral form within the condenser, and is hence termed the worm. The condenser is kept full of cold water from the tub at the left. By carefully regulating the temperature, one liquid may be separated from another by "fractional" distillation, advantage being taken of the fact that each liquid has its own boil-

---

* That freezing is a warming process may be conclusively shown as follows: Gently melt some sodium sulphate (a cheap salt that may be obtained from any apothecary) in a flask by heating it over a lamp flame. Put it aside to cool slowly in a perfectly quiet place. After cooling it remains liquid, but ready to freeze as soon as motion among its molecules is started. Disturb it by putting a thermometer bulb into the liquid. At once crystals are seen shooting out, and the mass is soon frozen hard. The mercury in the thermometer meanwhile rises, and the warming may be felt with the hand.

252   HEAT.

ing-point, higher or lower than that of the liquid with which it is mixed.

Boiling-point.—When we heat water, the bubbles which pass off first are the air dissolved in the liquid; next bubbles of steam form on the bottom and sides of the vessel, and, rising a little distance,

Fig. 182.

A Still.

are condensed by the cold water. Collapsing, they produce the sound known as "simmering." As the temperature of the water rises, they ascend higher, until they burst at the surface, and pass off into the air. The violent agitation of the water thus produced is termed boiling.* Some substances vaporize

---

* The temperature of water can not be raised above the boiling-point, unless the steam be confined. The extra energy is applied in expanding the water into steam. This occupies 1,700 times the space, and is of the same temperature as the water from which it is made. Nearly 1,000° units

at ordinary temperatures; others only at the highest; while the gases of the air are but the vapor of substances which boil at exceedingly low temperatures. The distinction between gases and vapors in ordinary language is only relative.

The boiling-point of water depends on three circumstances: (1.) *Purity of the water.* A solid substance dissolved in water ordinarily elevates the boiling-point. Thus salt water boils at a higher temperature than pure water. The air dissolved in water tends by its elastic force to separate the molecules. If this be removed, the boiling-point may be elevated to 275° F., when the water will be converted into steam with explosive violence.

(2.) *Nature of the vessel.* Water will boil at a lower temperature in iron than in glass. When the surface of the glass is chemically clean, the boiling-point is still higher. This seems to depend in some degree on the strength of the adhesion between the water and the containing vessel.

(3.) *Pressure upon the surface* raises the boiling-point.* Water, therefore, boils at a lower temperature on a mountain than in a valley. The temperature of boiling water at Quito is 194° F., and on

of heat for each pound of water are expended in this process, but are made sensible again as temperature when the steam is condensed. Steam is invisible. This we can verify by examining it where it issues from the spout of the tea-kettle. It soon condenses, however, into minute globules, which become visible in white clouds.

* Pressure opposes the repellent heat-force, and so renders it easier for cohesion to hold the particles together. In the interior of the earth there may be masses of matter heated red or white hot and yet solid, more rigid even than glass, in consequence of their melting-point being raised so high by the tremendous pressure that they can not liquefy.—TAIT.

Mont Blanc, 183° F. The variation is so uniform that the height of a place can thus be ascertained; an ascent of 596 feet producing a difference of 1° F.

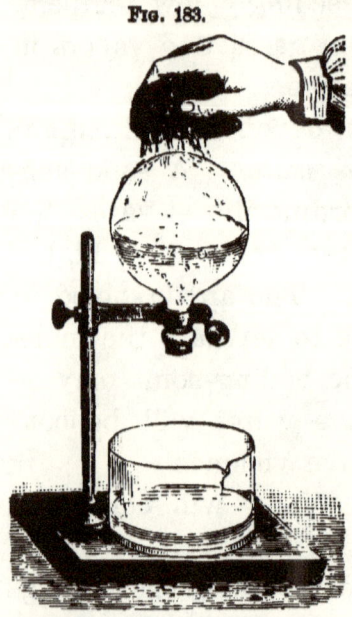

Fig. 183.

Boiling Water by Condensing its Vapor.

The influence of pressure is well illustrated by the following experiment: Half fill a strong glass flask with water, and boil this until all the air is expelled from both the water and the space above it. Now quickly apply a tight stopper and invert. The pressure of the steam will stop ebullition. A few drops of cold water will condense the steam, and boiling will recommence. This will soon be checked, but can be restored as before. The process may be repeated until the water cools to the ordinary temperature of the air, and even then the liquid inside may be made to boil by rubbing the outside of the flask with ice. The cushion of air which commonly breaks the fall of water is removed, and if the cork be air-tight, the water, when cold, will strike against the flask with a sharp, metallic sound.

6. Evaporation is a slow formation of vapor, which takes place at ordinary temperatures. Water evaporates even at the freezing-point. Clothes dry in the open air in the coldest weather. The wind

PHYSICAL EFFECTS OF HEAT. 255

quickens the process, because it drives away the moist air near the clothes and supplies dry air. Evaporation is also hastened by an increase of surface and a gentle heat.

Vacuum pans are employed in condensing milk and in the manufacture of sugar. They are so arranged that the air above the liquid in the vessel may be exhausted, and then the evaporation takes place rapidly, and at so low a temperature that burning is avoided.

The cooling effect of evaporation is due to the absorption of energy required to drive the molecules apart beyond their spheres of mutual attraction. Water may be frozen under the receiver of an air-pump by placing a small watch-glass containing it over a pan of strong sulphuric acid, which absorbs the vapor as fast as it is formed in the vacuum. The cooling due to rapid evaporation of a part is sufficient to freeze the rest. By strong pressure and cooling, carbonic acid is easily liquefied. Allowing a jet of this liquid to escape, the evaporation of a part of it causes the rest to freeze into a snowy powder which may be pressed into a ball.* Nitrogen, oxygen, and air, which is a mixture chiefly of these

---

* Mercury in contact with it is quickly solidified. On throwing the frozen metal into a little water, the mercury instantly liquefies, but the water turns to ice, the solid thus becoming a liquid and the liquid a solid by the exchange of heat. A cold knife cuts through the mass of frozen mercury as a hot knife would ordinarily through butter. The author, on one occasion, saw Tyndall, during a course of lectures at the Royal Institution at London, when freezing a ladle of mercury in a red-hot crucible, add some ether to hasten the evaporation. The liquid caught fire, but the metal was drawn out from the glowing crucible, through the midst of the flame, frozen into a solid mass.

two gases, have been liquefied. Liquid air boils at −337° F. *in a vacuum*. Nitrogen has been obtained in "snow-like crystals of remarkable size," and by reducing the pressure on these a temperature of −373° F. was attained,—the lowest recorded up to the present date (1887).

7. Spheroidal State.—If a few drops of water be put in a hot, bright spoon, they will gather in a globule, which will dart to and fro over the surface. It rests on a cushion of steam, while the currents of air drive it about. If the spoon cool, the water will lose its spheroidal form, and coming into contact with the metal, burst into steam with a slight explosion.*

8. Specific Heat.—More energy is required to raise the temperature of a pound of water through one degree than for any other substance except the gas hydrogen. The number of heat units (p. 249, § 3) required to raise one pound of a given substance one degree F. is called its *specific heat;* thus for mercury it is about $\frac{1}{30}$; for iron, $\frac{1}{8}$; for air, nearly $\frac{1}{4}$; for hydrogen, $3\frac{4}{10}$. On this account the ocean changes its temperature far less quickly than the land, and sea-side cities are subject to less extremes of temperature than those on the middle of a continent. On the elevated plateau region around the Great Salt Lake the temperature during the year varies from 115° F. to −30° F.

---

* Drops of water spilled on a hot stove illustrate the principle.—By moistening the finger, we can touch a hot flat-iron with impunity. The water assumes this state, and thus protects the flesh from injury.— Furnace-men can dip their moistened hands into molten iron.

III. COMMUNICATION OF HEAT.

Heat tends to become diffused equally among neighboring bodies.* There are three modes of distribution.

1. Conduction *is the process of heating by the passage of heat from molecule to molecule.*—*Example:* Hold one end of a poker in the fire, and the other end soon becomes hot enough to burn the hand. Of the ordinary metals, silver and copper are the best conductors.† Wood is a poor conductor, especially "across the grain."

Gases are the poorest conductors; hence porous bodies, as wool, fur, snow, charcoal, etc., which contain large quantities of air, are excellent non-conductors. Refrigerators and ice-houses have double walls, filled between with charcoal, sawdust, or other non-conducting substances. Air is so poor a conductor that persons have gone into ovens that were hot enough to cook meat, which they carried in and laid on the metal shelves; yet, so long as they did not themselves touch any good conductor, they experienced little inconvenience.

Liquids are also poor conductors.—*Example:*

---

* If we touch an object colder than we are, it abstracts heat from us, and we say "it feels cold"; if a warmer body, it imparts heat to us, and we say "it feels warm." Adjacent objects have, however, the same temperature, though flannel sheets feel warm, and linen cold. These effects depend upon the relative conducting power of different substances. Iron feels colder than feathers because it robs us faster of our heat.

† Place a silver, a German-silver, and an iron spoon in a dish of hot water. Notice how much sooner the handle of the silver spoon is heated than the others.

Hold the upper end of a test-tube of water in the flame of a lamp. The water nearest the blaze will boil without the heat being felt by the hand.

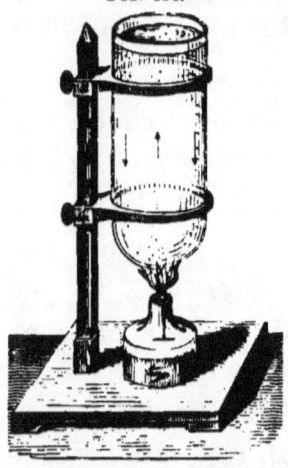

Fig. 184.

Heating by Convection.

2. Convection *is the process of heating by circulation.* (1.) CONVECTION OF LIQUIDS.—Place a little sawdust in a flask of water, and apply heat. We shall soon find that an ascending and a descending current are established. The water near the lamp becoming heated, expands and rises. The cold water above sinks to take its place.

(2.) OF GASES.—By testing with a lighted candle, we shall find at the bottom of a door opening into cold air, a current setting inward, and at the top, one setting outward. The cold air in a room flows to the stove along the floor, is heated, and then rises to the ceiling. Heating by hot-air furnaces depends upon the principle that warm air rises.

3. Radiation.—Two bodies not in contact may yet exchange heat across the intervening space. This transfer is effected by means of the ether of space, and is called *radiation*. A hot stove throws the ether surrounding it into vibrations, which are transmitted *radially* outward. On being absorbed (stopped) by bodies radiant energy is again converted into heat.* The

---

*The Radiometer is an instrument that, for a time, was supposed to exhibit the actual mechanical force of the sunbeam. It consists of a tiny

power of radiating heat and of absorbing radiant energy differs for different substances. In general, good radiators are also good absorbers, and *vice versa*. The space between the earth and the sun is not warmed by the sunbeam, because the ether of space is incapable of absorbing radiant energy. Dry air is so poor an absorber that meat can be cooked by radiation, while the surrounding air remains at the freezing-point. Aqueous vapor is a comparatively good absorber, especially for the longer waves. For this reason the moisture in the atmosphere stops much of the energy of the sunbeam passing through it, and converts it into heat. The greater part of the solar energy absorbed by the earth is again radiated forth into space, but in longer waves. To these glass is especially opaque.* Hence its use in the construction of green-houses. At lofty

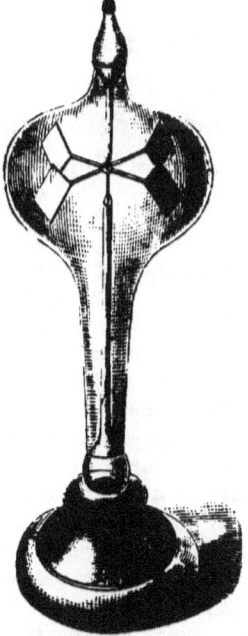

Fig. 185.

The Radiometer.

vane delicately pivoted in a glass globe from which the air is exhausted as fully as possible, when the globe is hermetically sealed. The four arms of the vane carry each a thin light disk of mica or aluminium, covered with lamp-black on one side and uncovered on the other. When daylight falls upon it the little vane revolves rapidly. The motion ceases as soon as the light is cut off. When different gases are admitted into the globe, the rate of rotation varies. It is now believed that the unequal heating of the black and white surfaces of the disks causes unequal reaction of the molecules of air left in the vacuum. Lamp-black being the best absorber and radiator, receives the more forcible bombardment of flying air molecules.

* In the course of Prof. Langley's experiments upon Mount Whitney, water was boiled by exposing it in a copper vessel covered by a pane of window-glass, to the direct rays of the sun. This

elevations the dry air allows the heat received by the soil during the day to escape so rapidly that a frost occurs before morning, though the heat of the afternoon may be torrid.

4. Absorption and Reflection.—A good reflector is a poor absorber and also a poor radiator. Snow is a good reflector, but a poor absorber or radiator. Light colors often absorb solar heat less and reflect more than dark colors.* White is generally considered the best reflector, and black the best absorber and radiator. But the nature of the material is of more importance than its tint. If on a bright summer day three thermometers are exposed to the sun, one held up in mid-air, another resting on a bed of black silk, and the third on a bed of white sand, it will be found in a short time that the temperatures indicated will be very different. The thermometer on the sand will have its bulb more warmed than that on the bed of black silk; and both of these will be warmer than the one in mid-air.

shows that many of the heat-rays of the sunbeam are stricken down by the air before reaching low levels, but may be utilized at high elevations. So, were the atmosphere removed the earth would receive far more heat and yet be much colder than now, because there would be no beds of water vapor to check the radiation back into space. See "American Journal of Science," March, 1883.

* Experiments show that with artificial heat the molecular condition of the surface varies radiation as well as reflection. In fact, white lead is as good a radiator as lamp-black.—On one side of a sheet of paper paste letters of gold-leaf. Spread over the opposite side a thin coating of scarlet iodide of mercury—a salt which turns yellow on the application of heat. Turn the scarlet side down. Hold over the paper a red-hot iron. The gold-leaf will reflect the heat, but the paper spaces between the letters will absorb it, and on turning the paper over, the gilt letters will be found traced in scarlet on a yellow background.

IV. THE STEAM-ENGINE.

WHEN steam rises from water at a temperature of 212°, it has an elastic force of nearly 15 lbs. per square inch. If the steam be confined and the temperature raised, the elastic force will be rapidly increased.

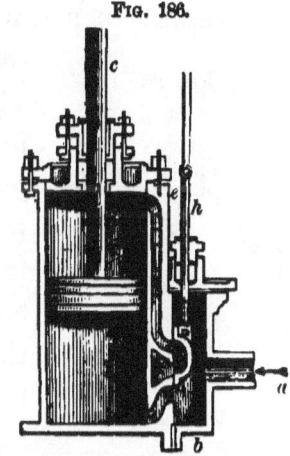

Fig. 186.

Steam-chest and Cylinder of an Engine.

1. The Steam-engine is a machine for using the elastic force of steam as a motive power. There are two classes, *high-pressure* and *low-pressure*. In the former, the steam, after it has done its work, is forced out into the air; in the latter, it is condensed in a separate chamber by a spray of cold water. As the steam is condensed in the low-pressure engine, a vacuum is formed behind the piston; while the piston of the high-pressure engine acts against the pressure of the air. The elastic force of the steam must be 15 lbs. per square inch greater in the latter case. The figure represents the piston and connecting pipes of an engine. The steam from the boiler passes through the pipe, a, into the steam-chest, b, as indicated by the arrow. The sliding-valve worked by the rod h lets the steam into the cylinder, alternately above and below the piston, which is thus made to play up and down by the expansive force. This valve is so arranged that at the moment fresh steam is let in on one side of the piston, the spent steam on the other side is released into the outer air, or into the condensing chamber.

2. The Governor is an apparatus for regulating the supply of steam. *AB* is the axis around which the heavy balls *E* and *D* revolve. They are so connected by hinge-joints that the ring at *B* may be pulled down or lifted by them, while that at *C* is fixed. When the machine is going too fast, the balls fly out and thus pull down the rod, *K*, which is in connection with a valve that controls the pipe supplying steam. A portion of the steam is thus obstructed, and the revolution of the balls becomes slower. This in turn makes them descend, and in so doing they lift the rod, *K*. The valve is thereby opened and more steam supplied, whenever the speed of revolution becomes too small.

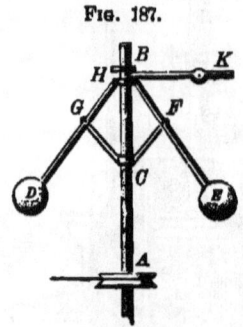

Fig. 187.

The Governor.

3. A High-pressure Engine is shown in Fig. 188. *A* represents the cylinder; *B*, the steam-chest at its side, connected with it on the interior by the sliding-valve already shown; *C*, the throttle-valve in the pipe through which steam is admitted from the boiler; *D*, the governor; *E*, the band-wheel by which the governor is driven; *F*, the pump; *G*, the crank; *I*, the conductor attached to *a*, the cross-head; *H*, the eccentric rod (*h* in Fig. 186) which works the sliding-valve in the steam-chest; *K*, the

High-pressure Steam-engine.

governor-valve; *S*, the shaft by which the power is conveyed to the machinery. The cross-head, *a*, slides to and fro in a groove, and is fastened to the rod which works the piston in the cylinder *A*. The expansive force of the steam is thus communicated to *a*, thence to *I*, by which the crank is turned. The heavy fly-wheel renders the motion uniform (p. 23).

V. METEOROLOGY.

1. General Principles.—(1.) The air always contains moisture. The amount it can receive depends upon the temperature; warm air absorbing more, and cold air less. At 100° F., a cubic foot of air can hold nearly 20 grains of invisible water vapor; a reduction of 70° will cause nine tenths of that quantity to be condensed into visible droplets. When the air at any temperature contains all the vapor it can hold in an invisible state, it is said to be saturated; *any fall of temperature will then condense a part of the vapor.*

(2.) When air expands against pressure (*i.e.*, doing work in the expansion), its energy, being thus expended, ceases to be manifested as temperature. The warm air from the earth ascending into the upper regions, is thus rarefied and cooled. Its vapor is then condensed into clouds, and often falls as rain. Owing to this expansion of the atmosphere and the greater radiation of heat in the dry air of the upper regions, there is a gradual diminution of the temperature as the altitude increases, the mean rate in the north temperate zone being about 1° for 300 feet.

2. Dew.*—The grass at night, becoming cooled by radiation, condenses the vapor of the adjacent air upon its surface. Dew will gather most freely upon the best radiators, as they will the soonest become cool. Thus grass, leaves, etc., receive the largest deposits. It will not form on windy nights, nor when there are clouds in the sky to reflect the heat radiated from the ground. In tropical regions the nocturnal radiation on clear nights is often so great as to render the formation of ice possible. In Bengal, water is exposed for this purpose in shallow earthen dishes resting on rice straw In parts of Chili, Arabia, etc., by its abundance, dew feebly supplies the place of rain. When the temperature of plants falls below 32°, the vapor is frozen upon them directly, and is called *white*, or *hoar-frost*.

3. Fogs are formed when the temperature of the air falls below the *dew-point, i. e.*, the temperature at which dew is deposited for a given degree of humidity. They are characteristic of low lands, rivers, etc., where the air is saturated with moisture.

---

* Dew was anciently thought to possess wonderful properties. Baths in this precious liquid were said greatly to conduce to beauty. It was collected for this purpose, and for the use of the alchemists in their weird experiments, by spreading fleeces of wool upon the ground. Laurens, a philosopher of the middle ages, claimed that dew is ethereal, so that if we should fill a lark's egg with it and lay it out in the sun, immediately on the rising of that luminary, the egg would fly off into the air!

266   HEAT.

4. Clouds differ from fogs only in their elevation in the atmosphere. They are produced chiefly by the cooling due to expansion as currents of warm moist air rise high above the surface of the ground. In tropical regions they float only at great heights; in arctic regions, near the ground.

FIG. 189.

Different kinds of Clouds—one bird indicates the Nimbus, two birds the Stratus, three birds the Cumulus, and four birds the Cirrus cloud.

The *stratus* cloud is composed of broad, widely-extended cloud-belts, sometimes spread over the whole sky. It is the lowest cloud, and often rests on the earth, where it forms a fog. It is the night-cloud.

The *cumulus* cloud is made up of large cloud-masses looking like snow-capped mountains piled up along the horizon. It forms the summits of pillars of vapor, which, streaming up from the earth, are condensed in the upper air. It is the day-cloud. When of small size and seen near midday, it is a sign of fair weather.

The *cirrus* (curl) cloud consists of light, fleecy clouds floating high in air. It is composed of little needles of ice or flakes of snow.

The *cirro-cumulus* is formed by small rounded portions of cirrus cloud, having a clear sky between. Sailors call this a "mackerel sky." It accompanies warm, dry weather.

The *cirro-stratus* is produced when the cirrus cloud spreads into long, slender strata. It forebodes rain or snow.

The *cumulo-stratus* is due to increase in thickness of the cumulus clouds, becoming denser and darker below, while the upper parts flatten out and thus appear like the stratus clouds. They often precede thunder-storms.

The *nimbus* cloud is that from which rain falls. It may be produced by the thickening of any of the forms just described.

5. Rain is the product of rapid condensation of vapor in the upper regions. At a low temperature the vapor is frozen directly into *snow*. This may melt before it reaches the earth, and fall as rain or sleet. A sudden draught of cold air into a heated ball-room has produced a miniature snow-storm. The wonderful variety and beauty of snow-crystals are illustrated in the figure.

Rain always warms the air. Vapor can not condense without giving out as heat to the surrounding medium the energy it absorbed in assuming the vaporous state.* It has been estimated that the heat given to the west coast of Ireland by rain-fall is equivalent

Fig. 190.

Snow-crystals.

to half of that derived from the sun. At Cherrapoonjee, in India, the annual rain-fall is four times as great as on the coast of Ireland.

6. Winds are produced by variations in the tem-

* "A gallon of water weighs ten pounds, and if spread out so as to form a layer an inch thick, it would cover about two square feet of space. To cover a square mile an inch in depth, 60,000 tons of rain are required, or 12,000,000 gallons. In the condensation of the vapor needed to produce a single gallon, heat enough is given out to melt 75 pounds of ice, or to make 45 pounds of cast-iron white-hot. An inch of rain-fall on each square mile hence implies an evolution of heat sufficient to melt a layer of ice spread over the ground 8 inches thick, or to liquefy a globe of iron 130 feet in diameter, or a rod of it a foot in thickness and 260 miles in length."

perature of the air. The atmosphere at some point is heated and expanded; it rises and colder air flows in to supply its place. This produces currents. The *land and sea breezes* of tropical islands are caused by the unequal specific heat of land and water. During the day the land becomes more highly heated than the water, and hence toward noon a sea-breeze sets in from the ocean, and is strongest in the afternoon. At night the land cools faster than the water, and so a land-breeze sets out from the land, and is strongest after midnight.

Trade-winds are so named because by their regularity they favor commerce. A vessel on the Atlantic Ocean will sometimes, without shifting a sail, set steadily before this wind from Cape Verde to the American coast. The air about the equator is highly heated, and, rising to the upper regions, flows off north and south. The cold air near the poles sets toward the equator to fill its place. If the earth were at rest this would make an upper current flowing from the equator, and a lower current flowing toward it. As the earth is rotating on its axis from west to east, the under current starting from the poles is constantly coming to a part moving faster than itself. It therefore lags behind. When it reaches the north equatorial regions, it lags so much that it becomes a current from the north-east, and in the south equatorial regions a current from the south-east.

7. Ocean Currents are produced in a similar manner. The water heated by the vertical sun of the tropics rises and flows toward the poles. The Gulf Stream carries the heat of the Caribbean Sea across the Northern Atlantic to the shores of Scotland and Norway. This great stream of warm water, flowing steadily through the cold water of the ocean, rescues England from the snows of Labrador. Were it not for the barrier of a chain of mountains connecting North and South America, Great Britain would be condemned to arctic glaciers.

8. Adaptations of Water.—The great specific heat of water exercises a marked influence on climate. It tends to prevent sudden changes of weather. In the summer it absorbs vast quantities of heat, which it gives off in the fall, and thus moderates the approach of winter. In the spring the melting ice and snow drink in the warmth of the sunbeam. Since so much heat is required to melt the ice and snow, they dissolve very slowly, and thus ward off the disastrous floods which would follow, if they passed quickly into the liquid state.

Water contains air, which is necessary for the support of animal life. This air not only makes it available as a home for fish and other creatures that inhabit the water, but also makes the change from water to steam more gradual. Much of it is driven off when the water is heated. When water has been carefully deprived of the air usually held in solution, it is liable to violent commotion at any moment when it is heated above 212° F. With such water every stove-boiler would need a thermometer. A tea-kettle would require as careful watching as a steam-engine, and our kitchens would witness frequent and perhaps disastrous explosions.

Water, like other liquids, expands when heated and contracts when cooled for any temperatures above 39.2° F. The density of water is greatest at this temperature. Cooled below 39.2° F. water expands till it becomes solid at 32° F. One volume at 39.2° F. becomes 1.00013 volumes at 32° F.* This expansion in cooling is probably due to the regrouping of the molecules of the water preparatory to crystallization, the crystalline structure requiring more space than the liquid form. As soon as a few crystals are definitely formed, each serves as a nucleus around which others gather, and the process becomes then far more rapid. Certain metals, such as bismuth and iron, act like water in this respect, and are hence well fitted for making sharp castings, filling every crevice of the mold as they expand in crystallizing.† This crystallization of water is of great importance in connection with the freezing of our lakes and rivers. Were it not crystalline when

---

* Since ice when it melts contracts, pressure aids in liquefaction and so lowers the melting-point. In descending over the rough surface of a mountain slope, glacier ice is subjected to alternate pressure and extension. The pressure melts it and makes the mass slide farther down. Passing over some ledge it snaps, producing great fissures. When the walls of these come into contact they freeze together again, but only to be re-melted.

† Fit a small flask with a cork, through which passes an upright glass tube. Fill with colored water. Apply heat to the flask until the liquid runs over the top of the tube. This shows the expansion by heat. Now apply a freezing mixture to the flask, and at first the liquid in the tube falls, but soon begins to rise. When it runs over as before, apply heat and it shrinks back again. Thus *cold will expand and heat contract it.* When water is at its maximum density (about 39°) expansion sets in alike, whether you heat or cool it.

frozen, the water at the surface during severe weather, radiating its heat and becoming chilled, would contract and fall to the bottom, while the warm water below would rise to the top. This process would continue until the freezing-point was reached, when the whole mass would solidify into ice. Our lakes and rivers would freeze solid every winter. This would be fatal to all animal life in it, at least of the higher orders, such as fish. In the spring the ice would not, as now, buoyant and light, float and melt in the direct sunbeam, but, lying at the bottom, would be protected by the non-conducting water above. The longest summer would not be sufficient to thaw the deeper bodies of water. As it is, the ice is formed at the surface, and there it floats, protecting the water beneath from further reduction of temperature.*

Water, in freezing, has a tendency to free itself from salts and other substances dissolved in it. Thus, melting ice furnishes a means of obtaining fresh water in Arctic regions. If a barrel of vinegar freeze, we shall find much of the acid collected in a mass about the center of the ice.

* Water distills from the ocean and land as vapor, at one time cooling and refreshing the air, at another moderating its wintry rigor. It condenses into clouds, which shield the earth from the direct rays of the sun, and protect against excessive radiation. It falls as rain, cleansing the air and quickening vegetation with renewed life. It descends as snow, and, like a coverlet, wraps the grass and tender buds in its protecting embrace. It bubbles up in springs, invigorating us with cooling, healing draughts in the sickly heat of summer. It purifies our system, dissolves our food, and keeps our joints supple. It flows to the ocean, fertilizing the soil, and floating the products of industry and toil to the markets of the world. (See "Chemistry," p. 56-63.)

PRACTICAL QUESTIONS.

1. Why will one's hand, on a frosty morning, freeze to a metallic door-knob sooner than to one of porcelain?
2. Why does a piece of bread toasting curl up on the side exposed to the fire?
3. Why do double windows protect better than single ones from the cold?
4. Why do furnace-men wear flannel shirts in summer to keep cool, and in winter to keep warm?
5. Why do we blow our hands to make them warm, and our soup to make it cool?
6. Why does snow protect the grass in winter?
7. Why does water "boil away" more rapidly on some days than on others?
8. What causes the crackling sound of a stove when a fire is lighted?
9. Why is the tone of a piano higher in a cold room than in a warm one?
10. Ought an inkstand to have a large or a small mouth?
11. Why is there a space left between the ends of the rails on a railroad track?
12. Why is a person liable to take cold when his clothes are damp?
13. What is the theory of corn-popping?
14. Could vacuum-pans be employed in cooking?
15. Why does the air feel so chilly in the spring, when snow and ice are melting?
16. Why, in freezing ice-cream, do we put the ice in a wooden vessel and the cream in a tin one?
17. Why does the temperature generally moderate when snow falls?
18. What causes the singing of a tea-kettle? *Ans.* The escaping steam is thrown into vibration by friction against the spout.
19. Why does sprinkling a floor with water cool the air?
20. How low a degree of temperature can be marked by a mercurial thermometer?
21. If the temperature is 70° F., what is it C.?
22. Will dew form on an iron bridge? On a plank walk?
23. Why will not corn pop when very dry?
24. When the interior of the earth is so hot, why do we get the coldest water from a deep well?
25. Ought the bottom of a tea-kettle to be polished?
26. Which boils the sooner, milk or water?
27. Is it economy to keep our stoves highly polished?
28. If a thermometer be held in a running stream, will it indicate the same temperature that it would in a pailful of the same water?
29. Which makes the better holder when one wishes to protect his hands from a hot dish, woolen or cotton?

30. Which will give out the more heat, a plain stove or one with ornamental designs?

31. Does dew fall?

32. What causes the "sweating" of a pitcher?

33. Why is evaporation hastened in a vacuum?

34. Does stirring the ground around plants aid in the deposition of dew?

35. Why does the snow at the foot of a tree melt sooner than that in the open field?

36. Why is the opening in a chimney made to decrease in size from bottom to top?

37. Will tea keep hot longer in a bright or a dull tea-pot?

38. What causes the snapping of wood when laid on the fire? *Ans.* The expansion of the air in the cells of the wood.

39. Why is one's breath visible on a cold day?

40. What gives the blue color to air? *Ans.* The particles floating in it reflect the blue light of the sunbeam.

41. How does the heat at two feet from the fire compare with that at four feet?

42. Why does the frost remain later in the morning upon some objects than upon others?

43. Is it economy to use green wood?

44. Why does not green wood snap?

45. Why will a piece of metal dropped into a glass or porcelain dish of boiling water increase the ebullition?

46. Which can be ignited the more quickly with a burning-glass, lampblack or white paper?

47. Why does the air feel colder on a windy day?

48. Could a burning-lens be made of ice?

49. Why is an iceberg frequently enveloped by a fog?

50. Would dew gather more freely on a rusty stove than on a bright kettle?

51. Why is a clear night colder than a cloudy one during the same season?

52. Why is no dew formed on cloudy nights?

53. Why will "fanning" cool the face?

54. How are safes made fire-proof?

55. Why can you heat water more quickly in a tin than a china cup?

56. Why will a woolen blanket keep ice from melting?

57. Does dew form under trees?

58. What is the principle of heating by steam?

59. What is the cause of "cloud-capped" mountains?

60. Show how the glass in a hot-house acts as a trap to catch the sunbeam.

61. Does the heat of the sun come in through our windows?

62. Does the heat of our stoves pass out in the same way?

63. The top of a mountain is nearer the sun, why is it not warmer?

SUMMARY.

64. What is hoar-frost? *Ans.* Frozen dew.
65. Why will a slight covering protect plants from frost? *Ans.* Because it prevents radiation.
66. Why is there no frost on cloudy nights? *Ans.* The clouds act like a blanket, to prevent radiation and keep the earth warm.
67. Can we find frost on the windows and on the stone flagging the same morning?
68. Why will not snow "pack" into balls except in mild weather?
69. Why is the sheet of zinc under a stove so apt to become puckered?
70. Why does a mist gather in the receiver of the air-pump as the air becomes rarefied?
71. Why are the tops of high mountains in the tropics covered with perpetual snow?

SUMMARY.

Heat is produced by longer and less refrangible waves and slower vibrations of ether than those which cause light. Solar energy may be radiated, reflected, refracted, absorbed, and polarized, whether manifested as light or heat. If we elevate the temperature of a body sufficiently, such as a piece of platinum, we can cause it to emit rays of both heat and light. A body which allows the radiant heat to pass through it easily is styled *diathermanous;* rock-salt is such a body, being to heat-rays what glass is to light-rays. The sun is the principal source of heat. But heat can be obtained by chemical and mechanical means. In burning coal we secure it by the former method. Mechanical energy may be changed directly into heat, as in striking fire with flint and steel, and in hammering a bullet on an anvil until it is hot. According to Joule's law, 772 feet fall of a given weight corresponds to 1° of rise of temperature in the same weight of water.

Among the physical effects of heat are a change of temperature, expansion, liquefaction, vaporization, and evaporation. The heat-force increases the kinetic energy of the molecules, thus elevating the temperature; and the increased vibration of the molecules causes an expansion of the body. The latter is so uniform in certain substances, such as mercury, that it is used to indicate changes of temperature, as in the thermometer. The expansion of the metals by heat is turned to account in many art processes. The walls of a gallery in the Conservatoire des Arts et Metiers in Paris, had begun to bulge. To remedy this, iron rods were passed across the building and screwed into plates on the outside of the walls. By heating the bars, they were expanded, when they were screwed up tightly. Being then allowed to cool, they contracted, thus drawing the walls back toward a perpendicular. The same has been done for weakened walls in many other places.

Heat is the great antagonist of cohesion. The liquid and gaseous states of bodies depend on its relative presence or absence (absolute cold is as yet only a theoretical condition, all bodies with which we are familiar being relatively warm). When the heat-force nearly balances that of cohesion, the body breaks down into a liquid, and when the repellent fairly triumphs, the particles fly off as a gas. Immediately before and after each of these marked changes, viz., of a solid to a liquid and of a liquid to a gas, the thermometer indicates a constant temperature. Thus water from melting ice affects the thermometer just as the ice does, and steam is no hotter than the boiling water. The heat which, in these processes, becomes hidden from the thermometer is called latent, though we now know that, having been occupied in doing internal work, it has merely become potential, and can be readily turned again into kinetic energy. The so-called latent heat of water is only the potential heat-energy of the separated molecules, which will reappear the instant the molecules collapse and come once more within the grasp of cohesion. On this principle is based the method of heating by steam. Evaporation is a slow change to vapor that takes place at all temperatures, but may be greatly increased by a diminution of pressure, as in a vacuum. It is a cooling process, and is practically applied to the manufacture of ice.

By the subtraction of heat, *i. e.*, by cold, and by the addition of pressure, which antagonizes the repellent heat-force, gases may be liquefied and even congealed, the transparent carbonic-acid gas thus becoming a snowy solid. What were formerly called the "permanent gases" (oxygen, hydrogen, etc.), have been liquefied by means of the cold produced by their rarefaction when they were suddenly released from a pressure of two or three hundred atmospheres.

Heat is *conducted* from molecule to molecule of a body, *radiated* in straight lines through air (or space), and *circulated* by the transference of heated masses through a change of specific gravity due to expansion. The first method is characteristic of solids, and the third of liquids and gases. The elastic force of steam increases when it is confined and a higher temperature is reached. The steam-engine utilizes this principle. There are two forms of this machine, the high-pressure and the low-pressure, according as the waste steam is ejected into the air or condensed in a separate chamber. The phenomena of dew, rain, etc., depend upon the fact that a change from a higher to a lower temperature causes the air to deposit its moisture.

HISTORICAL SKETCH.

DEMOCRITUS, the originator of the Atomic Theory, held that heat consists of minute spherical particles radiated rapidly enough to penetrate every substance. Until very recently, heat and light were thus reckoned among the Imponderables, *i. e.*, matter which has no weight. Aristotle considered heat more a condition than a substance. Bacon, in his "Novum Organum," wrote: "Heat is a motion of expansion." Locke, half a century later, said: "Heat is a very brisk agitation of the insensible parts of an object, which produces in us the sensation from whence we denominate the object hot, so that what in our sensation is heat, in the object is motion."

The material view, however, held its ground. At the beginning of the 18th century, Stahl elaborated a theory that a buoyant substance called *phlogiston* is the principle of heat, and that when a body burns, its phlogiston escapes as fire. In 1760, Dr. Black investigated and made known the principles of what he termed *latent* heat, *i. e.*, heat which becomes hidden when ice is turned into water or water into steam. Priestley discovered, in 1774, and Lavoisier afterward developed, the modern view of combustion. But the latter philosopher then advanced the theory that heat (caloric) is an actual substance, which passes freely from one body to another and combines at pleasure. Toward the close of the 18th century, Benjamin Thompson, better known as Count Rumford, a native of Woburn, Mass., but in the employ of the Elector of Bavaria, proved the convertibility of force. "He first took the subject," as Professor Youmans well remarks, "out of the domain of metaphysics, where it had been speculated upon since the time of Aristotle, and placed it on the true basis of physical experiment."

Soon the scientific world seemed to be ripe for this discovery, and it appears to have sprung up spontaneously in men's thoughts every-where. Mayer, a physician of Germany, and Grove, of England, proved the mutual relation of the forces, the latter first using the term "Correlation of Forces," since changed to Conservation of Energy. Joule discovered the law of the "Mechanical Equivalent of Heat," about 1843. In his famous experiments, he used pound-weights made to fall through a measured distance. Cords were attached to them, so that, as they fell, they turned a paddle-wheel placed in a box of water. Other liquids were used instead of the water. The rise of temperature in the liquids was carefully marked. The loss by friction in the apparatus was estimated, and so, at last, the dynamical theory of heat was fully demonstrated. Names of philosophers well known to us, such as Henry, Helmholtz, Faraday, Thomson, Maxwell, Le Conte, Youmans, Stewart, and Tyndall, are associated with the final establishment of this theory.

Consult, on this interesting subject, Tait's "Recent Advances in Physical Science"; Stewart's "Treatise on Heat"; Tyndall's "Heat a Mode of Motion"; Maxwell's "Theory of Heat"; Thurston's "History of the Growth of the Steam-engine"; Buckley's "Short History of Natural Science"; Smiles' "Lives of Boulton and Watt"; Youmans' "Correlation of the Physical Forces"; "Read and the Steam-engine"; "American Cyclopedia," Art. "Steam-engine"; "Popular Science Monthly," Vol. XII., p. 616, Art. "Liquefaction of Gases"; Scott's "Meteorology," and Thomson's "Cruise of the Challenger."

IX.
MAGNETISM.

"THAT power which, like a potent spirit, guides
The sea-wide wanderers over distant tides,
Inspiring confidence where'er they roam,
By indicating still the pathway home;—
Through Nature, quickened by the solar beam,
Invests each atom with a force supreme,
Directs the cavern'd crystal in its birth,
And frames the mightiest mountains of the earth
Each leaf and flower by its strong law restrains
And binds the monarch Man within its mystic chains."
                                              HUNT.

ANALYSIS OF MAGNETISM.

MAGNETISM.
1. MAGNETS.
2. THE MAGNETIC MERIDIAN.
3. LAWS OF MAGNETISM.
4. INDUCTION.
5. HOW TO MAKE A MAGNET.
6. THE COMPASS.
7. LINES OF FORCE.
8. POLARITY OF THE NEEDLE.
9. TERRESTRIAL MAGNETISM.

MAGNETISM.

1. Magnets.*—A natural magnet is an ore òf iron (Fe_3O_4, "Popular Chemistry," p. 156), generally known

Fig. 191.

Magnet Dipped in Iron Filings

as lodestone (Saxon, laedan, to lead), which has the power of attracting iron.† The artificial magnet is a

* The term is derived from the fact that an ore of iron possessing this property was first found at Magnesia, in Asia Minor.
† A few other elements, such as nickel and cobalt, are attracted

282 MAGNETISM.

steel bar that has acquired properties like those of lodestone. If it be straight, it is called a bar magnet; if U-shaped, a horseshoe magnet. A piece of soft iron called the *armature* is placed so as to connect the two ends of the horseshoe.

If we insert a magnet in iron filings, they will cling chiefly to its ends termed the *poles*. The magnetic force will be exerted even through any intervening body that is not itself magnetic.

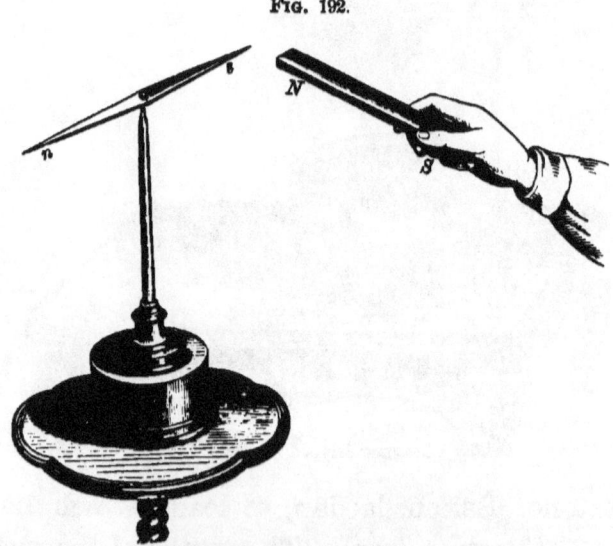

Fig. 192.

Influence of one Magnet on another.

2. The Magnetic Meridian.—If a slender bar magnet be suspended so as to swing freely in a horizontal plane, it will come to rest in a definite position; one pole pointing north, the other south. A vertical plane passing through the two poles in this

---

slightly by the magnet. They are all called magnetic bodies, but for ordinary purposes iron may be regarded as the only one of importance.

position is called the *magnetic meridian*. The pole pointing north is called the north or (+) pole, the other the south or (—) pole.

3. Laws of Magnetism.—If we hold a magnet near a magnetic needle, we shall find that the south pole of one attracts the north pole, and repels the south pole of the other.* This proves the law—
"*Like poles repel, and unlike poles attract.*"

Two opposite poles placed near together attract each other strongly, but this force, like gravitation, *diminishes as the square of the distance increases*.

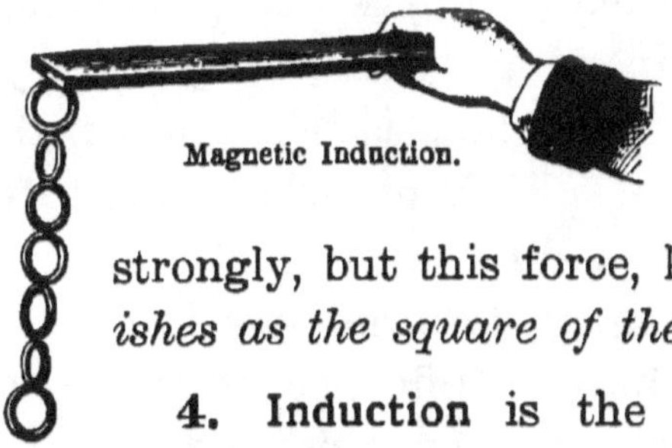

FIG. 193.

Magnetic Induction.

4. Induction is the process of developing magnetism by bringing a magnetic body and a magnet near together. If a piece of soft iron be brought near a magnet, it immediately assumes the magnetic state, but loses it on being removed. In steel the

---

* *Experiments.*—1. Rub the point of a sewing-needle across the north pole of a magnet. Bring the point near the south pole of the magnetic needle. The needle will be repelled, showing that the point of the sewing-needle has become a south pole. 2. Suspend a key from the north pole of a magnet. Bring the south pole of an equal magnet close to the upper end of the key. The key will instantly fall. 3. Suspend a long iron wire from the north pole of a magnet. Bring the north pole of the second magnet near the lower end of the wire. The wire is repelled, because its lower extremity possesses north polarity. 4. Immerse the unlike poles of two magnets in iron filings. Bring the two poles near each other. The filings will move toward one another. But if the poles of the magnets are like, the filings will fall off the magnets. 5. To ascertain whether a metallic substance contains iron: Bring the substance near one of the extremities of a magnetic needle. If the position of the needle be affected, then the substance almost certainly contains iron. A piece of copper will not affect the magnetic needle.

change is induced and lost much more slowly. The end of the bar next to the south pole of the magnet becomes the north pole of the new magnet, and *vice versa*. When opposite states are thus developed in the opposite ends of a body, it is said to be *polarized*. Whenever an object is attracted by a magnet, it is supposed first to be made a magnet (polarized) by induction, and then the attraction consists in that of unlike poles for each other. Thus we may suspend from a magnet a chain of rings held together by magnetic attraction.* Each link is a magnet with its north and south poles. Each particle of the tuft of filings in Fig. 191 is a distinct magnet. By inducing magnetism, a magnet does not lose force. It rather gains by the reciprocal influence of the new magnet. An armature acts in this manner to strengthen a magnet. If we break a magnet, the smallest fragment will have a north and a south pole. This is explained by supposing that every molecule contains two opposite kinds of energy which neutralize each other. When the bar is magnetized these are separated, but do not leave the molecule. This is hence polarized, the halves assuming opposite magnetic states, as shown in Fig. 194. The light half of each little circle represents the positive, and the dark the negative side. All the molecules exert their negative force in one direction, and their positive in

Fig. 194.

Polarization of an Iron Bar.

---

* Repeat this experiment with keys, or nails of different sizes, or bits of wire of varying length.

the other. The forces thus neutralize each other at the center, but manifest themselves at the ends of the magnet. Hence it is impossible to produce a magnet with only one pole. Each pole necessitates the presence of the other.

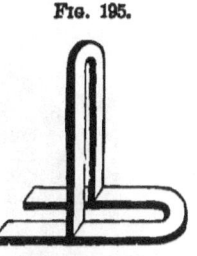

Fig. 195.

Making a Magnet.

5. How to Make a Magnet.—Place the inducing magnet, as shown in Fig. 195, on the unmagnetized bar (which any blacksmith can make from a bar of steel), and draw it from right to left several times, always carrying it back through the air to the starting-point.*

6. The Compass is a magnetic needle used by mariners, surveyors, etc. It is delicately poised over a card on which the "points of the compass" are marked. At most places the needle does not point directly N. and S. The *"line of no variation"* in the United States passes near Wilmington, N. C., Charlottesville, Va., and Pittsburg, Pa. East of this the declination of the needle from true north is toward the west, and west of it the declination is toward the east.

7. Lines of Force.—The region in the neighborhood of a magnet pole within which it can have a perceptible effect upon magnetic bodies is called its

---

* A needle may be magnetized by laying it across the poles of a horseshoe magnet. After remaining a few moments, the end in contact with the north pole of the magnet will become a south pole and the other a north pole. If it be suspended from the middle by a thread it will point north and south. A knife-blade may be magnetized by rubbing it several times, in the same direction, across one of the poles of the magnet.

286   MAGNETISM.

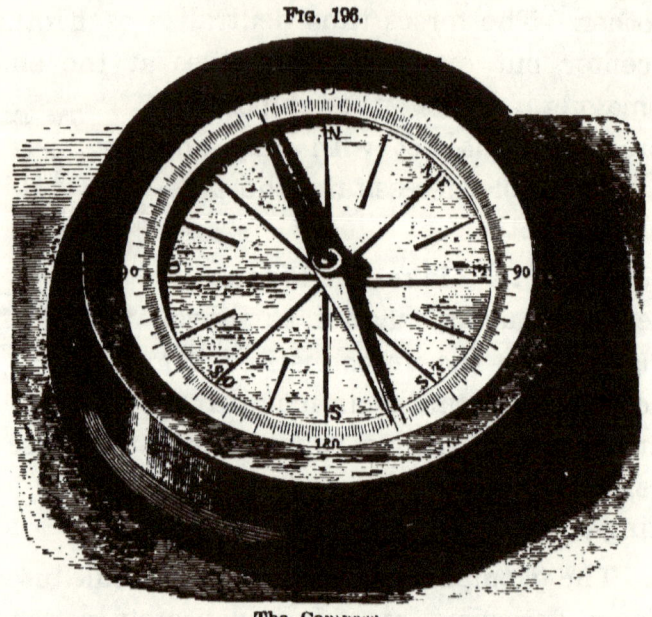

Fig. 196.

The Compass.

field. The directions which these bodies tend to assume are called *lines of force.* Over a bar magnet lay a sheet of paper or a plate of glass, and sprinkle fine iron filings over this. On gently tapping the

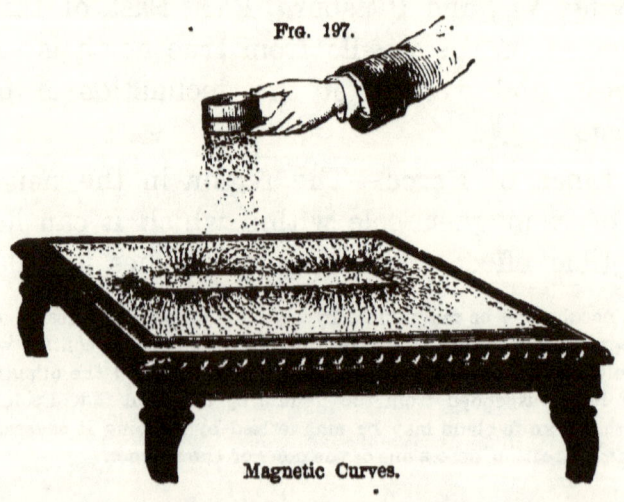

Fig. 197.

Magnetic Curves.

plate they become arranged in curved lines, many of which seem to radiate from the poles. These are the indicators of the lines of force, the position of each iron filing being determined by its direction and distance from the two poles of the magnet.

Between the two opposite poles of a horseshoe magnet, or of two separate magnets brought near together, the lines of force are straight.

8. Polarity of the Needle.—WHY THE NEEDLE POINTS NORTH AND SOUTH.—The earth is a great magnet, whose opposite poles produce lines of force that permeate its body and the space around. A needle when magnetized tends to assume the direction of the line of force that passes through it. The position of the terrestrial magnetic poles is not constant, and hence the needle changes its direction accordingly.

Suppose a magnet *NS* passing through the center of a small globe.

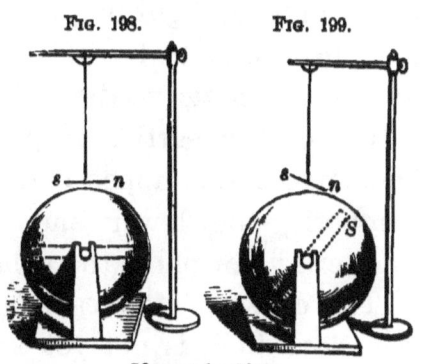

FIG. 198.   FIG. 199.

Magnet in Globe.

The needle *sn* will hang parallel to it (Fig. 198), its positive pole being attracted by the negative pole of the magnet, and *vice versa*. If the globe be revolved (Fig. 199), the positive pole of the needle will turn—*dip*, as it is termed—downward. If the globe be revolved in the other direction, the negative pole of the needle will dip in the same manner.*

* Similar phenomena are noticed in the compass. At the magnetic

A DIPPING-NEEDLE is poised as shown in Fig. 200. At the magnetic equator it hangs horizontally, but when carried north its positive end dips downward more and more until it points vertically downward at a point in Boothia Peninsula, north of Hudson Bay, called the *pole of verticity*, or often simply the magnetic pole. This is the *negative* pole of the earth. The position of its positive pole has been calculated to be beneath the Antarctic Ocean.*

Fig. 200.
The Dipping-needle.

9. Magnetism of the Earth.—Magnetic bodies such as iron fences, lightning-rods, iron standards of chairs, etc., are found to be fully magnetic when examined with a magnetic needle. This is due to the inductive action of the earth's magnetism. In the northern hemisphere the upper end of such bodies is south-magnetic, the lower end north-magnetic. In the southern hemisphere the polarity is the reverse of this.

The cause of the earth's magnetism and of the variations in it is not yet known.

equator it is horizontal, but *dips* whenever taken north or south. An unmagnetized needle, if poised, in our latitude, on being magnetized, settles down, as if the north end were the heavier. This is remedied by making the north end of the needle lighter, or by attaching a little weight upon the south end. The reverse is true in the southern hemisphere.

* The declination and dip of the magnetic needle have daily and yearly variation, and also very slow changes requiring centuries to complete. In 1686, at New York, the declination was 9° west; in 1750, 6° 20' west; in 1790, 4° 15' west; in 1847, 6° 30' west; in 1885, 8° west. The line of no variation was becoming slowly shifted eastward from 1686 to 1790, then became stationary, and has since been moving westward. The intensity of terrestrial magnetism at any given place has also daily variations, growing stronger by day and weaker by night.

SUMMARY.

Natural magnets are found in certain regions, but in practice, magnets made of steel are generally used. These bars may be magnetized either by contact with other magnets or by placing them within the magnetic field of a coil conducting an electric current (see p. 332). The existence of magnetism is manifested by polarity. Like poles repel, unlike attract each other. The intensity of the force varies inversely as the square of the distance. A magnet induces magnetism in any neighboring magnetic body. This is not prevented by intervening bodies which are not themselves magnetic. If free to move, small bodies thus influenced by induction tend to place themselves in certain directions, called lines of force, around the inducing magnet. The declination, the dip, and the intensity are the magnetic elements of a place. Each of these is subject to daily variations, and additionally to slow changes requiring many years for a cycle. The cause of the earth's magnetism is unknown. Sudden variations in it accompany the outbreak of spots on the sun, and magnetic storms are usually attended by the appearance of the aurora.

HISTORICAL SKETCH.

Magnets probably became first known to European nations through the discovery of natural magnets by the Greeks in the Thessalian district of Magnesia. From this the name was taken. The tendency of a magnetized needle to point in a definite direction was early noticed, and it is thought that the compass was invented by the Chinese. The first mention of the use of the magnetic needle in Europe occurs in 1190. The needle was floated on a cork, and in this way it served as a guide to the Chinese travelers. By the end of the fifteenth century the compass was known to most European sailors, and its use was specially frequent among the Spanish and Portuguese. The declination of the needle was known to the Chinese in the beginning of the twelfth century. Columbus discovered it independently in 1482, just ten years before his discovery of America. The first known work for the use of seamen was written during the reign of Queen Elizabeth. It was entitled "A Discourse on the Variation of the Cumpas or Magneticall Needle," and is dedicated to "the travaillers, sea-men, and mariners of England." The dip was discovered accidentally in 1576 by Robert Norman, an English instrument-maker. He found the dip at London to be 71° 50′. Dr. Gilbert, the physician of Queen Elizabeth, published his great work, "De Magnete," about 1600. In this he announces his belief that the earth is a great magnet, controlling the direction of the needle. The variation in intensity of the earth's magnetic force has become known chiefly during the present century.

X.

Electricity.

"MAXWELL has demonstrated that luminous vibrations can be nothing else than periodic variations of electro-magnetic forces. Hertz, in proving by experiments that electro-magnetic oscillations are propagated like light, has given an experimental basis to the theory of Maxwell. This gave birth to the idea that the luminiferous ether and the seat of electric and magnetic forces are one and the same thing. This being established I can now . . . reply to the question you put to me: What is electricity? It is not only the formidable agent which now and then shatters and tears the atmosphere, terrifying you with the crash of its thunder, but it is also the life-giving agent which sends from heaven to earth, with light and heat, the magic of colors and the breath of life. It is that which makes your heart beat to the palpitations of the outside world; it is that which has the power to transmit to your soul the enchantment of a look and the grace of a smile."—PROF. FERRARIS.

ANALYSIS OF ELECTRICITY.

ELECTRICITY.

I. Frictional Electricity.
1. Development of Electricity.
2. The Electroscope.
3. Electrical Attraction and Repulsion.
4. Theory of Electricity.
5. Electrical Potential.
6. Electrical Conduction.
7. Electrical Density.
8. Electrical Induction.
9. The Plate Machine.
10. Theory of Attraction.
11. Free and Bound Charges.
12. Inductive Capacity.
13. Electrical Condensation.
14. The Leyden Jar.
15. The Voss Machine.
16. Lightning.
17. Effects of Frictional Electricity.

II. Voltaic Electricity.
1. Simple Voltaic Cell.
2. Action in the Voltaic Cell.
3. The Electric Current.
4. The Volt.
5. Electrical Resistance.
6. The Ohm.
7. The Ampère.
8. The Battery.
9. Polarization within the Battery.
10. Special Forms of Battery.
11. Effects of Voltaic Electricity.

III. Transformations of Electric Energy.
1. Effect of a Voltaic Current on a Magnetic Needle.
2. Magnetic Coils.
3. The Tangent Galvanometer.
4. The Electro-magnet.
5. The Electro-magnetic Telegraph.
6. Ocean Cables.
7. Current Induction.
8. The Induction Coil.
9. Magneto-induction.
10. The Telephone.
11. The Microphone.
12. The Magneto-electric Machine.
13. The Dynamo-electric Machine.
14. The Electric Light.
15. Thermo-electricity.
16. Animal Electricity.

ELECTRICITY.

When a piece of sealing-wax is rubbed with flannel it *temporarily* acquires the *new* property of attracting, when brought near them, other light bodies, such as shreds of paper, gold leaf, lint, etc. If the hand is approached sufficiently near the excited sealing-wax, small sparks may not infrequently be seen in the dark to leap across the space between the hand and the sealing-wax.* The excited sealing-wax is evidently in a different state from what it was before it was rubbed with the flannel. The supposed agent producing this state is called *electricity*, and the sealing-wax in this excited condition is said to be *electrified*. Other substances may be used instead of the sealing-wax. A piece of vulcanite, glass, resin, shellac, or amber will do as well. The substance also with which friction is produced may be varied. Instead of flannel, fur, silk, or chamois-skin covered with tin-amalgam may be used, and it is found that with some rubbers better results are obtained than with others. Under proper conditions most bodies may be rendered electric by friction.

* In cold, frosty weather, a person, by shuffling about in his stocking-feet upon the carpet, can develop so much electricity in his body that he can ignite a jet of gas by simply applying his finger to it.—Blasts in mines intended to be fired by electricity have thus been prematurely discharged by the workmen touching the wires. To prevent this disastrous effect, at tne Sutro Tunnel, Nevada City, the workmen who are handling exploders wet their boots, stand on an iron plate to conduct off the electricity of the body, and wear rubber gloves.

I. FRICTIONAL ELECTRICITY.

1. Electricity developed either directly or indirectly by friction is called *frictional electricity*. For its detection it is convenient to employ instruments specially adapted for this purpose, called electroscopes.

2. The Electroscope.—A pith ball suspended by a silk thread, as shown in Fig. 201, constitutes a simple and very effective electroscope. A straw 3 or 4 inches

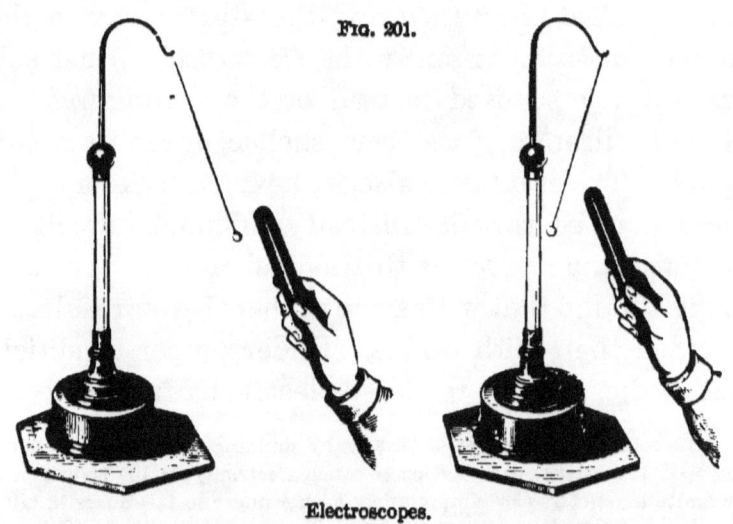

Fig. 201.

Electroscopes.

in length, and suspended by a silk fibre at the middle so as to hang horizontally, may be substituted for the pith ball. A lath balanced on an egg placed in a wineglass may also serve as an electroscope.

3. Electrical Attraction and Repulsion.—If a warm dry glass, such as a lamp-chimney, be rubbed with a silk handkerchief, a crackling sound will be heard.

If the tube be held near the face, a sensation like that of touching cobwebs will be felt. Make with the electrified tube the following experiments.* Present

* The following simple experiments are instructive:—1. A rubber comb passed a few times through the hair will furnish enough electricity to turn the lath entirely around, and empty egg-shells, paper hoops, etc., will follow the comb over the table in the liveliest way.—2. Take a thin sheet of gutta-percha, about a foot square; lay it upon the table, and rub it briskly a few times with an old fur cuff; the gutta-percha will become powerfully electrified.—3. Lift the gutta-percha by one corner, and some force will be required to separate it from the table.—4. Hold the electrified gutta-percha in the left hand; bring the fingers of the right near the paper; it will be attracted to the hand, and sparks will pass to the fingers with a snapping sound.—5. Hold some feathers, suspended by a silk thread, near the excited gutta-percha, and the feathers will be attracted.—6. Hold the excited paper, or the excited sheet of gutta-percha, over the head of a person with dry hair; the hair will be attracted by the gutta-percha, and each particular hair will stand on end.—7. Hold the excited gutta-percha near the wall; the gutta-percha will fly to it, and remain some minutes without falling.—8. Place a sheet of gutta-percha on a tea-tray; rub the gutta-percha briskly with a fur cuff; place the tea-tray with the excited sheet of gutta-percha on a dry tumbler; lift off the gutta-percha from the tea-tray; bring the knuckle of your hand near the tray, and you will receive a spark. Replace the gutta-percha on the tray and apply your knuckle, and you will receive another spark. This may be repeated a dozen times.—9. Take a sheet of foolscap paper and a board about the same size. Heat both till they are thoroughly dry. While hot, lay the paper on the board and rub the former briskly with a piece of rubber. The paper and board will cling together. Tear the paper loose and try experiments 4, 5, 6, and 7. Return the paper and rub as before. Cut the paper so as to form a tassel. Then lift, and the strips of the tassel will repel one another.—10. Take a piece of common brown paper, about the size of an octavo book, hold it before the fire till quite dry and hot, then draw it briskly under the arm several times, so as to rub it on both sides at once by the coat. The paper will be found so powerfully electrical, that if placed against a wainscoted or papered wall of a room, it will remain there for some minutes without falling.—11. While the paper still clings to the wall hold against it a light, fleecy feather, and it will be attracted to the paper in the same way the paper is to the wall.—12. If the paper be warmed, drawn under the arm as before, and then hung up by a thread attached to one corner, it will sustain several feathers on each side; should these fall off from different sides at the same time, they will cling together very strongly; and if after a minute they are all shaken off, they will fly to one another in a singular manner.—13. Warm and excite the paper as before, and then lay on it a ball of elder-pith,

it to the pith ball of an electroscrope. This will be attracted till it touches, and then fly off. The end of the suspended straw will likewise be first attracted, but then repelled just after it is touched. Grasp the pith ball or straw for a moment. It will no longer be repelled. Rub a stick of sealing-wax with a woolen cloth or some fur. The behavior of the pith ball or straw toward it will be the same as toward the glass. But bring the rubbed sealing-wax near to the pith ball or straw that is repelled by the rubbed glass; there will be attraction instead of repulsion. If the excited glass be held on one side of a ball and the

about the size of a pea; the ball will immediately roll across the paper, and if a needle be pointed toward it, it will again roll to another part, and so on for a considerable time.—14. Support a pane of glass, well dried and warmed, upon two books, one at each end, and place some bran underneath; then rub the upper side of the glass with a silk handkerchief, or a piece of flannel, and the bran will dance up and down like the images in Fig. 208. —15. Place a common tea-tray on a dry, clean tumbler. Then take a sheet of foolscap writing-paper (as in No. 9) and dry it carefully until all its hygrometric moisture is expelled. Holding one end of the sheet on a table with the finger and thumb, rub the paper with a large piece of India rubber a dozen times vigorously from left to right, beginning at the top. Now take up the sheet by two of the corners and bring it over the tray, and it will fall like a stone. This, as well as the apparatus in No. 8, forms a simple *Electrophorus*, fit to perform many experiments ordinarily performed with that instrument. If the tip of a finger be held close to the bottom of the tray, a sensible shock will be felt. Next, lay a needle on the tray with its point projecting outward, remove the paper, and, in the dark, a star sign of the negative electricity will be seen; return the paper, and the positive brush will appear. Lay a dry, hot board, as in No. 9, on top of four tumblers. If a boy stand on the board he will be insulated, and on his holding the tray vertically, the paper will not fall. Sparks may then be drawn from his body, and his hair will be electrified.—16. Warm a lamp-chimney, rub it with a hot flannel, and then bring a downy feather near it. On the first moment of contact, the feather will adhere to the glass, but soon after will fly rapidly away, and you may drive it about the room by holding the glass between it and the surrounding objects; should it, however, come in contact with any thing not under the influence of electricity, it will instantly fly back to the glass.

excited wax on the other, it will fly between the two, touching each in succession alternately. From this we conclude that (1), there are *two kinds of manifestation of frictional electricity;* and (2), *like kinds are manifested by repulsion, and unlike by attraction.* The electricity from the glass is termed positive [+], and that from the wax, negative [—].*

4. Theory of Electricity.—It is thought that positive and negative electricity exist in every body, in a state of total or partial equilibrium. When this is disturbed, as by friction, electrical separation follows, and each kind becomes manifested, just as in the polarization of a magnet, if the proper conditions are observed. Electricity is not a fluid, as was long taught. It may be a condition of strain among the molecules of a body, capable of being communicated like a fluid. We know only its laws, and not its nature.

5. Electric Potential.—A body electrically excited by friction or otherwise is said to be *charged.* The charge may be either positive or negative, strong or weak. If two bodies equally and oppositely charged are put into contact, the charge of each is neutralized by that of the other. A body strongly charged positively is said to be at high potential; if nega-

---

* In the following list, each substance becomes positively electrified when rubbed with the body following it; but negatively, with the one preceding it.—GANOT.

1. Cat's fur.	5. Cotton.	9. Shellac.	13. Caoutchouc.
2. Flannel.	6. Silk.	10. Resin.	14. Gutta-percha.
3. Ivory.	7. The hand.	11. The metals.	15. Gun-cotton.
4. Glass.	8. Wood.	12. Sulphur.	

tively, at low potential; when discharged, at zero potential. The surface of the earth is electrically at zero.

6. Conductors and Insulators.—A body which allows electricity to pass through it freely is termed a *conductor;* one which does not, is called a *bad conductor*, or *insulator*. Copper is one of the best conductors, and hence it is used in many electrical experiments. Glass is one of the best insulators. A body is said to be insulated when it is supported by some bad conductor, which is generally glass or vulcanite. A body can be highly charged only when insulated. In damp air electricity is quickly dissipated. This is due to the deposit, on the glass insulators, of a thin film of moisture, which conducts away the electricity. For success in electrical experiments, therefore, it is important to keep the air dry and warm, since dry air is one of the best of insulators.*

7. Distribution of Electricity on Bodies.—A charge communicated to one part of an insulator is not spread over its whole surface; but when a good conductor is charged at any point the spread is instantaneous. It spreads, however, *only* on the surface, and

* The following list contains some of the most common conductors and insulators:

Conductors.		Insulators.	
Metals.	Vegetables.	Dry Air.	Glass.
Charcoal.	Animals.	Shellac.	Silk.
Flame.	Linen.	Amber.	Dry Paper.
Minerals.	Cotton.	Sulphur.	Caoutchouc.
Acids.	Dry Wood.	Wax.	
Water.	Ice.		

FRICTIONAL ELECTRICITY. 299

not through the interior. A pith ball, if made to touch the outside of an electrified metal cup or hol-

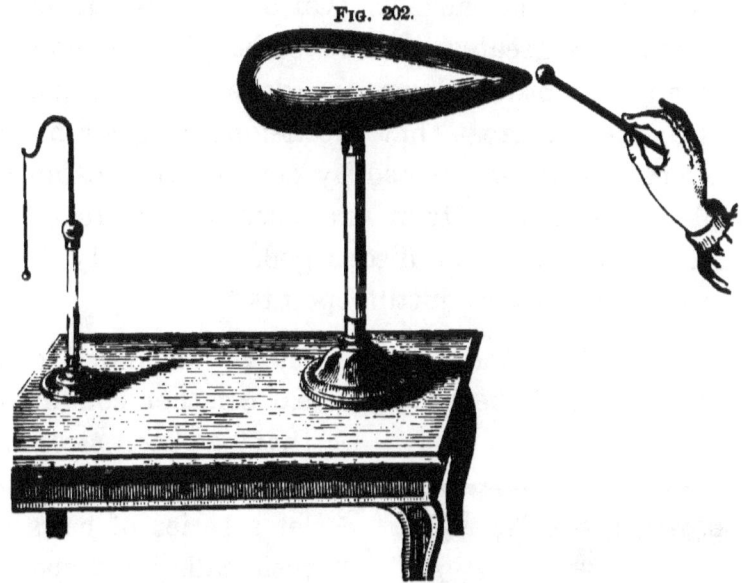

Fig. 202.

Variation in Electric Density.

low ball, is strongly repelled; but on the interior there is no such effect.* If the ball is spherical, the

Fig. 203.

Faraday's Conical Bag.

* Faraday once made a hollow cube of wood, measuring 12 ft. each way and covered with tin-foil. Insulating this, he charged it with a powerful machine until sparks darted off from every corner on the outside. Going within this little room with his most delicate electroscopes, he could not detect the least effect upon them. He made a conical bag of linen, and fastened its open end to an insulated ring. Pulling it out with a silken cord, he electrified it. The charge was manifest on the outside, zero on the inside. Reversing the pull so as to turn it inside out, the new exterior was found to be charged. A half-minute previously it had been a neutral interior. The student should try this interesting experiment, using the most delicate electroscope that he can make.

amount of electricity at all points of its surface is the same; or, we may say that the *electric density* is uniform over its surface. On a cylinder the electric density is greatest at the ends. If one end is blunt and the other sharp, the density at the sharp end becomes so great that the neighboring air molecules are quickly electrified by contact and instantly repelled. Others in turn are successively repelled, and the body is soon discharged. Electricity thus escapes rapidly from jutting points.*

8. Electrical Induction. — Let an insulated conductor, Fig. 204, be brought near another conductor that has been strongly charged positively, and let a series of pairs of pith balls be suspended from the first. The motion of the balls shows that the ends of the insulated conductor are electrically excited, while the middle is neutral. The end nearest the charged conductor is excited negatively and the remote end positively. If the charged conductor be removed, all of the pith balls collapse. Place several insulated

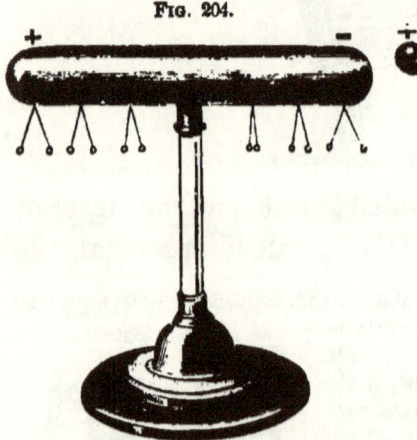

Fig. 204.

Electrical Induction.

---

* The *electric whirl*, mounted on the prime conductor of an electrical machine, illustrates this action. As each molecule of air is repelled from a point, it reacts with equal force against the point. This is sufficient to set the light wire-wheel in rapid rotation.

FRICTIONAL ELECTRICITY. 301

conductors, as shown in Fig. 205, the balls being strongly charged, that at the right positively, and that at the left negatively. Each intermediate conductor becomes excited, as indicated, and becomes

FIG. 205.

Electrical Induction.

neutral when the balls are discharged. It has been *polarized by induction*, like a magnetic body when brought into the field of a magnet pole.*

9. The Plate Electrical Machine consists of (1) a circular glass *plate* which can be turned by means of a crank; (2) a pair of leather or cloth *rubbers* pressed against the plate and covered with electrical amalgam or tin disulphide; † (3) a metallic *comb* or fork with sharp points which nearly touch the plate; (4) a *prime conductor*, consisting of a rounded brass cylinder, insulated by resting on a glass standard, and connected at one end with the comb. Frequently

---

* The experiment in Fig. 204 can be nicely performed by means of an egg placed flatwise on the top of a dry wine-glass and the glass tube represented in Fig. 201. Several eggs and glasses will show the principle of Fig. 205. See Tyndall's "Lessons in Electricity," p. 39.

† Electrical amalgam is a mixture of two parts each of tin and zinc, and four parts of mercury. By experience it has been found that, when this is rubbed on glass, electrical separation is most easily effected. Tin disulphide is often called "mosaic gold," because of its metallic yellow color. It is used in bronzing.

the lower half of the plate is made to revolve between a pair of silken flaps (Fig. 206). A chain is usually attached to the knob in connection with the rubber, and connects this with the ground through the medium of a gas-pipe or other conductor.

On turning the crank, the friction of the plate against the rubbers produces electrical separation;

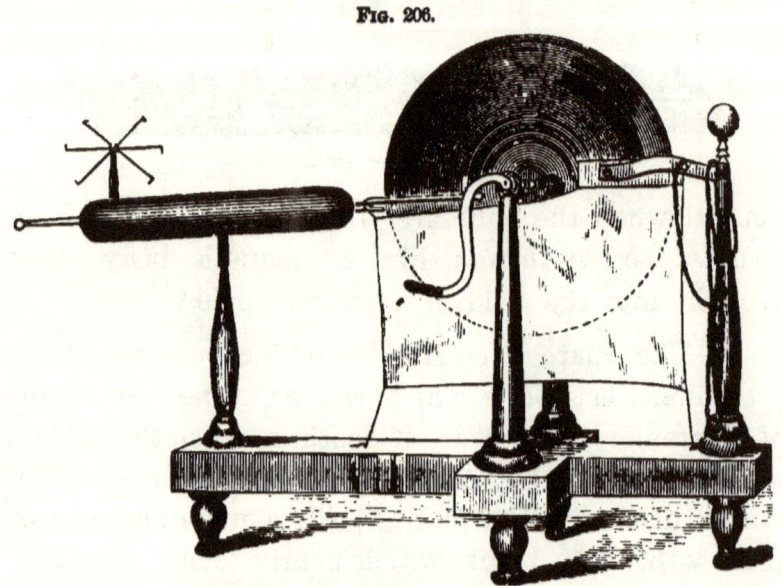

FIG. 206.

The Plate Electrical Machine.

the rubbers becoming charged negatively, the plate positively. The negative charge is conducted off to the earth by the chain, which thus restores the rubbers to zero potential. The positive charge on the plate, when this is brought opposite the comb, polarizes the prime conductor and comb by induction. Positive electricity becomes manifested on the remote conductor, and negative electricity at the comb is communicated at once by the sharp points to the air, whose molecules are repelled into contact with the plate, thus neutralizing its positive charge. The prime conductor is hence left charged to high potential.

The action of the plate machine is thus an application of both friction and induction.

10. The Theory of Attraction is likewise an application of induction. In Fig. 201, where a glass rod at high potential is brought near a pith ball, this is polarized by induction, the nearer half becoming negative, and the remote half positive. The charge on the rod attracts the negative half and repels the positive half. But since the negative half is nearer, the attraction exceeds the repulsion, and the pith ball moves toward the rod. On touching this the negative charge is wholly neutralized, and only repulsion can be effective. Every case of electrical attraction is thus a case of induction.

The *electric chime* consists of three bells, two of which, c and b, are hung by brass chains, while the middle one is insulated above by a silk cord, and connected below with the earth by a chain. The balls between them are also insulated. The outer bells becoming charged with positive electricity from the prime conductor of an electrical machine, polarize the balls by induction

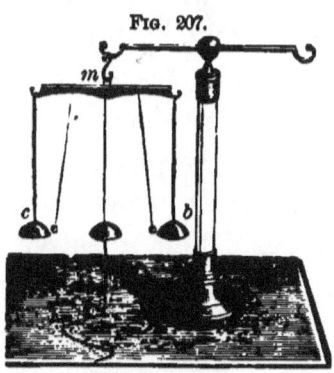

Fig. 207.

Electric Chimes.

through the intervening air. The balls being then attracted to the bells, are charged and immediately repelled. Swinging away, they strike against the middle bell, discharging their electricity, and are forthwith attracted again. Flying to and fro, they ring out a merry song.

Fig. 208.

Dancing Images.

The *dancing image* consists of a pith-ball figure placed between two metallic plates, the upper one hanging from the prime conductor, and the lower one connected with the earth. The dance is conducted by alternate attraction and repulsion.*

11. Free and Bound Electricity.—The gold-leaf electroscope is more sensitive than one of pith balls. Within a dry glass jar a pair of strips of gold-leaf are suspended from a metal rod terminating at the top in a knob or plate. If a rod, excited for example negatively, be brought near the knob, then by induction this becomes charged positively while both leaves are charged negatively, and hence repel each other. By placing the finger on the knob and withdrawing it while the rod is still near, the leaves collapse. Their negative charge has been conducted off to the earth. But on withdrawing now the rod, they diverge again and remain apart. The positive charge

---

* A slow motion should be given to the electrical wheel, and a pin thrust into the heel of the image will add much to the stamp of the tiny feet.

on the knob was *"bound"* there by the presence of the negatively excited rod and could not be conducted away, like the negative charge on the leaves. On removing the rod after the finger has been taken away the positive charge becomes *"free";* it is distributed over knob and leaves, and these now repel each other with a positive charge. So long as a charge is "bound," *i.e.*, so long as the knob and leaves under the influence of the excited rod are at zero potential, it fails to manifest itself.

Gold-leaf Electroscope.

12. Inductive Capacity.— A body through which induction occurs is called a *dielectric*. The power with which inductive influence acts across a dielectric is called its *inductive capacity*, and this is different for different dielectrics. Induction takes place better through a plate of glass than through a layer of equal thickness of shellac or air. In other words, glass has a higher inductive capacity than shellac or air, and is, therefore, the best dielectric of the three.

13. Electrical Condensation.— By putting a good

dielectric between two conducting surfaces, one of which is connected with the prime conductor of an electrical machine and the other with the earth, electricity may be strongly "condensed" on these surfaces. In Fig. 210, let the strip on the left represent a conductor, of tin-foil, positively charged from the machine, and that on the right a similar conductor connected with the earth, the intervening space being occupied by a plate of glass. This dielectric becomes polarized, the surface on the right attaining a negative charge which is bound there, while the corresponding positive electricity on the same side is neutralized by connection with the earth. The negative charge in turn reacts through the dielectric, binding a positive charge on the left, whose energy thus becomes potential. The conductor can then receive a new charge from the machine, and the process is repeated until the greatest charge is accumulated that the condenser can carry. Its molecules are then in a condition of great strain.

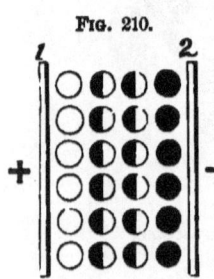

FIG. 210.

14. The Leyden Jar consists of a glass jar, serving as dielectric, coated inside and outside, not quite to the top, with tin-foil. It is fitted with a cover of baked wood through which passes a metal rod with a knob at the top, and below a metal chain extending down to the inner coating. The jar is *charged* by bringing the knob near the prime conductor of the machine, while the outer coating communicates with the earth. The inner coating becomes charged first from the machine, a succession of sparks being received until the two coatings acquire a large charge of bound electricity, positive within and negative without. To *discharge* it * one end of a conductor with an insulated handle is put on the outer coating, while the other is brought near the knob above. A sharp snap and a brilliant flash through the air announce that equilibrium is restored. Minute particles detached from the solid conductors are made momentarily white-hot, giving brilliancy to the spark. †

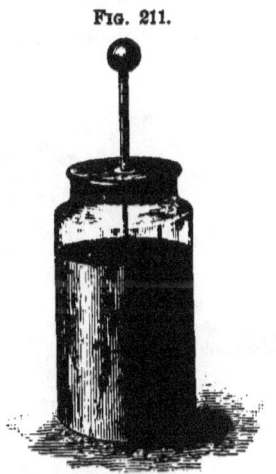

Fig. 211.

The Leyden Jar.

The tin-foil on a Leyden jar serves only as a conductor, and not as an accumulator, of the charge. The jar may be made with movable coatings. After it is charged these may be removed. Putting the same jar then into another set of coatings, it may be discharged in the usual manner.

---

* It is said that Cuneus, a pupil at Leyden, discovered the principle of the Leyden jar in the following curious way: While experimenting, he held a bottle of water to the prime conductor of his electrical machine. Holding the bottle with one hand, he happened to touch the water with the other, when he received a shock so unexpected, and so unlike any thing he had ever felt before, that he was filled with astonishment. It was two days before he recovered from his fright. A few days afterward, in a letter to a friend, the physicist innocently remarked, that he would not take another shock for the whole kingdom of France.

† The incredibly small quantity of the metal volatized in this way is a striking proof of the divisibility of matter. During some experiments at the Philadelphia mint a gold pole lost in weight by a strong spark one millionth of a grain; and $\frac{1}{7140000}$ of a grain of nickel *signed its name* in the spectroscope brilliantly. See " Popular Science Monthly," May, 1877.

15. The Toepler-Holtz Electrical Machine.—Many
improvements have been made on the plate electrical machine. One of the best is the Toepler-Holtz machine.* This consists of a fixed glass plate in front of which re-

Fig. 212.

The Toepler-Holtz Machine.

volves a smaller one provided with six metallic buttons (Fig. 212, b). On the rear of the fixed plate are two sheets of varnished paper, a and a', called the armature. On each a strip of tinfoil is cemented, from which a metallic arm extends around to the front, ending in a rubber of brass filaments, r. Under this each but-

* The "Wimshurst" electrical machine is the most recent improvement of electrical machines depending on induction. In these machines electricity is generated by glass discs faced with metallic sectors, which are rotated in opposite directions. These machines are certain in their action, and furnish large quantities of electricity at a high potential.

ton passes. A pair of combs, c and c', connect with adjustable discharging rods, p and n. Another pair of combs and brushes are attached to the brass rod, dd'; this extends across in front of the plate, which revolves in the direction shown by the arrows.

If there be the least possible difference of potential between the two armatures, such as is naturally due to accidental conditions on their surfaces, it may be greatly increased by revolving the plate. Suppose the left armature, a', to be faintly charged positively, while the right armature, a, is neutral; then a' induces a slight negative bound charge on the button in front, which in revolving passes under d'. Passing from d' to r, the button comes opposite a neutral armature. Its negative bound charge at once becomes free and is conducted through the rubber r to the armature behind, charging it negatively. This at once acts inductively on the button, causing it to acquire a positive bound charge with which it passes d. This charge is freed at r' and conducted to the armature a', strengthening its positive charge. This process continues, both armatures becoming soon strongly and oppositely charged. The comb, c', by induction from a, is polarized. It discharges negative electricity, while the rod, p, acquires a strong positive charge. In like manner c discharges positive electricity and n acquires a strong negative charge. A succession of sparks soon passes between p and n, the strength of which is greatly increased by condensation in the Leyden jars, with which the discharging rods are connected.

16. Lightning is only the discharge of a Leyden jar on a grand scale. If two clouds with opposite charges of electricity come near together, the intervening air reaches its limit of polarization, and a flash occurs like that between the discharging-rods of an electrical machine.* The air is never quite uniform in conducting power at all places, and the immense spark, moving along the line of least resistance, describes a zigzag course. It suddenly heats the air, which expands and instantly collapses. The concussion produces a series of air-waves from successive parts of the spark. These constitute thunder, which continues to roll because the sound is reflected many times from clouds, and from masses of air which differ among themselves in density. Often the charged cloud approaches the ground rather than another cloud. Discharge takes place, and exposed objects, such as tall houses or trees, are destroyed if included in the lightning's path.

Lightning-rods were invented by Franklin.† They are based on the principle that electricity always

---

* The air is constantly electrified. In clear weather it is in a positive state, but in foul weather it changes rapidly from positive to negative, and *vice versa*. Dr. Livingstone tells us that in South Africa the hot wind which blows over the desert is so highly electrified, that a bunch of ostrich feathers held for a few seconds against it becomes as strongly charged as if attached to an electrical machine, and will clasp the hand with a sharp, crackling sound.

† Franklin's plan was opposed by many men of his day, who declared it was as impious to ward off Heaven's lightning, "as for a child to ward off the chastening rod of its father." There was much discussion as to whether the conductors should be pointed or not. Wilson persuaded George III. that the points were a republican device to injure His Majesty, as they would certainly "invite" the lightning, and so the points on the lightning-rods upon Buckingham Palace were changed for balls.

seeks the best conductor. The rod should be pointed at the top with some metal which will not easily corrode. If constructed in several parts, they should be securely jointed. The lower end should extend into water, or else deep into the damp ground, beyond a possibility of any drought rendering the earth about it a non-conductor, and be packed about with ashes or charcoal. If the rod is of iron, it needs to be much larger than one of copper, which is a better conductor. Every elevated portion of the building should be protected by a separate rod. Chimneys need especial care, because of the ascending column of vapor and smoke. Water conductors, tin roofs, etc., should be connected with the damp ground or the lightning-rod, that they may aid in conveying off the electricity.*

DURATION OF THE FLASH.—The duration of the flash from a Leyden jar has been found to vary from two thousandths to forty billionths of a second. When the plate of the Toepler-Holtz machine is revolving at the highest speed, each button can be momentarily seen, as if it were still, when illuminated by the spark. The trees swept by the tempest, or a train of cars in rapid motion, when seen by a flash of

---

* The value of a lightning-rod consists, most of all, in its power of quietly restoring the equilibrium between the earth and the clouds. By erecting lightning-rods, we thus lessen the liability of a sudden discharge. Every drop of rain, and every snow-flake, falls charged with electric energy, and thus quietly disarms the clouds of their terror. The balls of electric light, called by sailors "*St. Elmo's fire*," which sometimes cling to the masts and shrouds of vessels and the flames said to play about the points of bayonets, indicate the quiet escape of electricity from the earth toward the clouds.

lightning, seem motionless; while a cannon-ball, in swift flight, appears poised in mid-air.

17. Effects of Frictional Electricity.—(1.) PHYSICAL.—Discharges from a large battery of Leyden jars will melt metal rods, perforate glass, split wood, magnetize steel bars, etc.—Let a person stand upon an insulated stool and become charged from the prime conductor. His hair, through repulsion, will stand erect in a ludicrous manner. On presenting his hand to a little ether contained in a warm spoon, a spark leaping from his extended finger will ignite

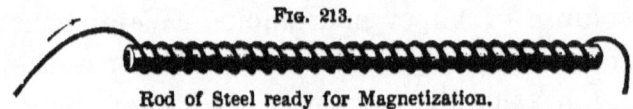

FIG. 213.
Rod of Steel ready for Magnetization.

it. If he hold in his hand an icicle, the spark will readily dart from it to the liquid.*—A card held between the knob of a Leyden jar and that of the discharger, will be punctured by the spark.—A piece of steel may be magnetized by the discharge from a Leyden jar. Wind a covered copper wire around a steel bar, as in Fig. 213, or inclose a needle in a small glass tube, around which the wire may be wound. On passing the spark through the wire, the needle will attract iron filings.—When strips of tinfoil are pasted on glass, and figures of various patterns cut from them, the electric spark leaping from

---

* This experiment can be more surely performed by using disulphide of carbon. The insulating stool may be merely a board laid on four dry flint-glass bottles or goblets, and the electricity be developed by rubbing a glass tube.

VOLTAIC ELECTRICITY. 313

one to the other presents a beautiful appearance.—If a battery be discharged through a small wire the electricity will be changed to heat, and the wire, if sufficiently small, will be fused into globules or dissipated in smoke.

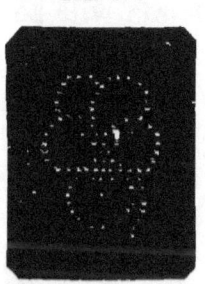

Fig. 214.

Illuminated Pane.

(2.) CHEMICAL EFFECTS.—The "electric gun" is filled with a mixture of oxygen and hydrogen gases. A spark causes them to combine with a loud explosion and form water.—The sulphurous smell which accompanies the working of an electrical machine, and is noticed in places struck by lightning, is owing to the production of ozone, a peculiar form of the oxygen of the air. (See "Popular Chemistry," p. 23.)

(3.) PHYSIOLOGICAL EFFECTS.—A slight charge from a Leyden jar produces a contraction of the muscles and a spasmodic sensation in the wrist. A stronger one becomes painful and even dangerous.

―――――•·•―――――

II. VOLTAIC ELECTRICITY.*

1. Simple Voltaic Circuit.—If a strip of zinc, coated over with mercury, be put into a mixture of sulphuric acid and water, no perceptible chemical action will be noticed. But if a strip of copper or

* This name is given in honor of the Italian physicist who made the first discoveries in this branch of electricity.

platinum be immersed at the same time, and the upper ends of the two pieces of metal be touched together or connected by wires, many little bubbles of gas will be seen on the second strip, forming and rising to the surface. When the experiment is performed in the dark, an almost infinitesimal spark is perceived at the moment the wires are joined.*

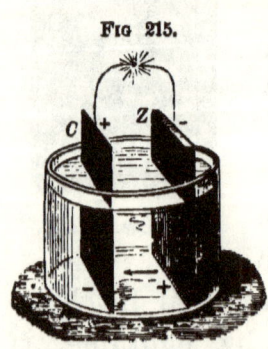

Fig 215.

A Voltaic Pair.

Two metal plates joined in this way form a *voltaic pair*. The exposed end of the copper or platinum plate is called the positive pole, and that of the zinc the negative pole of the pair.† Joining the wires, or otherwise connecting the poles, is termed *closing the circuit;* and separating them, *breaking the circuit*. A cup prepared for such an experiment is called a *voltaic cell*.

2. Action in the Voltaic Cell.—Zinc is far more easily acted upon by sulphuric acid than copper is. Each molecule of the acid is composed of two atoms of hydrogen, one of sulphur, and four of oxygen.‡ This may be expressed by the symbol, H_2SO_4. When the acid, mixed with water, acts on zinc (Zn), its hydrogen is set free, and a new substance, called zinc

---

* We can easily form a simple galvanic circuit by placing a silver coin between our teeth and upper lip, and a piece of zinc under our tongue. On pressing the edges of the two metals together, a peculiar taste will be perceived.

† These names may easily be remembered if we associate the p's with copper and positive, and the n's with zinc and negative.

‡ For the properties of these elements, refer to "Popular Chemistry," pp. 11, 38, and 103.

sulphate, is produced. Its symbol is $ZnSO_4$. It is at once dissolved in the water, leaving a fresh surface of metal to be attacked by the acid. It is thought that each molecule of liquid between the copper and zinc becomes polarized, then decomposed, giving up its H_2 to its neighbor on the side toward the copper, and its SO_4 to its neighbor on the side toward the zinc. The H_2 liberated, in contact with the copper, gathers in bubbles of gas; the SO_4, in contact with the zinc, unites with this metal to form zinc-sulphate, which dissolves in the liquid of the cell. If the exposed ends of the two plates be examined with a sufficiently delicate electroscope, while they are still separate, it is found that the copper is electrically at higher potential than the zinc. When they are connected, neutralization instantly takes place, but the action of the acid on the zinc renews the difference of potential between the copper and the zinc, so that the process continues as long as this action continues.*

3. The Electric Current.—The term "current" is applied to this continuous neutralization and renewal of difference of potential in the closed voltaic circuit. The current is said to "flow" through the conducting wire from the copper at high potential toward the zinc at low potential, just as water flows from an

* With what inconceivable rapidity must these successive changes take place in an iron wire to transmit the electric energy, as in actual experiments, from Valentia, Ireland, across the bed of the Atlantic and the American continent to San Francisco and return, a distance of 14,000 miles, in two minutes! In fact, it far surpassed the velocity of the earth's rotation, by which we measure time and leaving Valentia at 7:21 A. M., Feb. 1, it reached San Francisco at 11:20 P. M., Jan. 31.

elevated reservoir through a pipe toward a lower reservoir. There is no actual transfer of matter, no current of fluid; but only by analogy we may call it a current of energy transmitted through the entire thickness and length of the conducting wire. By analogy also the current is said to pass through the cell from zinc to copper, thus completing its circuit.

4. The Volt.—We measure the difference of temperature between two bodies in *degrees* on the thermometer scale, or the difference of level between the surfaces of two reservoirs in feet or meters. These are the accepted units of measurement. In like manner, for measuring the difference of potential between two bodies, a unit called the *volt** has been selected. In the simple voltaic cell already described, when freshly set in action, the difference of potential between its poles is about one volt. The force due to difference of potential is called *electro-motive force*, and is always measured in volts.†

* Named in honor of Alessandro Volta, an Italian physicist, who was born in 1745. In 1793, he communicated to the Royal Society of London an account of his important experiments on which the modern science of electricity has been largely built.

† The difference of potential between the discharging rods of a Voss electrical machine when giving long sparks is often several hundreds of volts. When passed through the body, such momentary currents are painful. The potential of the air during a thunder-storm quickly changes through thousands of volts. The voltaic cell furnishes a current that is exceedingly steady in comparison with the stream of sparks from an electrical machine, but of only small electro-motive force. Frictional electricity is sudden, noisy, convulsive; voltaic is gentle, silent, yet powerful. The one is like a quick, violent blow, as of a swiftly moving bullet; the other like the steady uniform pressure produced by a large mass slowly advancing against resistance. Lightning leaps across miles of air; the voltaic current will pass through a conductor from England to California rather than spark across half an inch of air. The most powerful frictional machine

5. Electrical Resistance.— Every conductor opposes resistance to the electric current. The amount of resistance offered depends upon the nature of the conductor, its length and cross-section. Metals offer much less resistance than liquids, and these again less than substances classed as insulators (§7, p. 298). For the same material the resistance increases with its length and decreases with increase of its cross-section.

6. The Ohm.—To measure resistance, a unit called the *ohm** has been selected. A piece of common copper wire, as thick as the band shown in Fig. 216, and fifty yards long, opposes a resistance of about one ohm. Coils whose resistance in ohms is known are much used in electrical measurement.

FIG. 216.

7. The Ampère.—To measure the effective current strength obtained from a voltaic cell, a unit called an *ampère* † has been selected. It is the

would be insufficient for telegraphing; while signals have been sent across the ocean with a tiny battery composed of "a gun-cap and a strip of zinc, excited by a drop of water the bulk of a tear." "Faraday immersed a voltaic pair, composed of a wire of platinum and one of zinc, in a solution of four ounces of water and one drop of oil of vitriol. In three seconds this produced as great a deviation of the galvonometer needle as was obtained by 30 turns of the powerful plate-glass machine. If this had been concentrated in one millionth of a second, the duration of an electric spark, it would have been sufficient to kill a cat; yet it would require 800,000 such discharges to decompose a grain of water."

* Named in honor of Dr. G. S. Ohm, a German physicist, who determined the relation existing between current strength, electro-motive force, and resistance.

† Named for André Marie Ampère, a French physicist, born in 1775, whose splendid work in electricity was such as to give him the highest rank along with Volta.

The definitions of electrical units given in the text are not exact enough to furnish more than the most elementary ideas. The student will

amount of current obtained when one volt of electro-motive force acts against one ohm of resistance.

With these units electricity can be measured with as much exactness as we measure quantities of grain or water.

8. A Battery consists of two or more voltaic cells so connected as to secure a stronger current than can be obtained from a single cell. According to Ohm's Law, the current strength (C) in ampères is equal to the electro-motive force (E) in volts divided by the resistance in ohms. The resistance is partly in the external conductor (R) and partly in the liquid of the battery (r). The law is expressed in a formula, thus,

$$C = \frac{E}{R + r}.$$

With a large number of cells a battery, therefore, can be arranged either to overcome a large external resistance, or, when this is small, to furnish a strong current.

9. Polarization within the Battery.—As soon as the action of a battery is well begun, the electro-motive force becomes rapidly diminished because hydrogen tends to collect upon the plate in connection with the positive pole. The bubbles interfere with further action and start a counter electro-motive force which neutralizes much of that in operation. Many different devices have been em-

find them more accurately defined in any text-book devoted specially to Electricity, such as that of Thompson or Urbanitzky.

ployed to diminish this evil, and each gives rise to a special kind of battery. Only a few need be described.

10. The Potassium Bichromate Battery.—Instead of copper, a pair of plates of carbon are immersed, with a plate of zinc between them, arranged so as to slide into the liquid or out of it at will. A solution of potassium bichromate in sulphuric acid and water is used. The sulphuric acid acts on the zinc, and the hydrogen is prevented from forming in bubbles by being combined at once with some of the oxygen which the chromic acid yields.

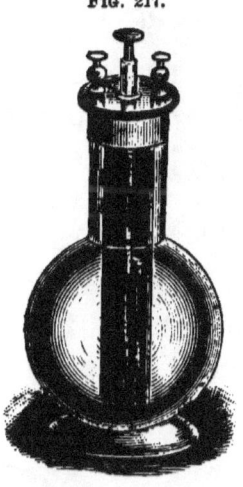

Fig. 217.

Potassium Bichromate Cell.

DANIELL'S BATTERY.— In this battery there are two fluids separated by a cup of porous earthenware, which does not prevent the passage of the current. In the outer vessel of glass there is a strong solution of copper sulphate ($CuSO_4$) in which a split copper cylinder is immersed. Within this is placed the porous cup, containing a rod of zinc coated with mercury and a mixture of sulphuric acid with water. Zinc sulphate is produced, and the liberated hydrogen decomposes some of the copper sulphate, taking its SO_4 and causing a deposit of metal-

Fig. 218.

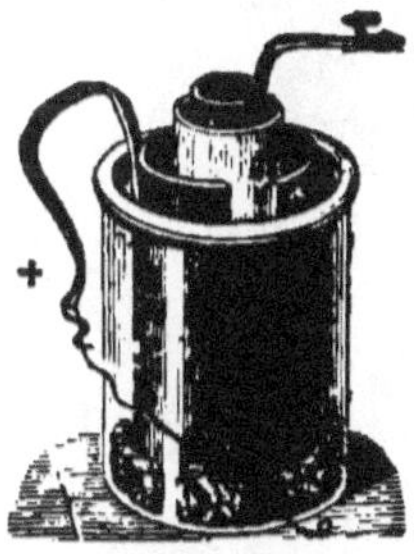

Daniell Cell.

lic copper. Polarization is thus prevented, and this is one of the most constant batteries known.

GROVE'S BATTERY.—In this the outer cup contains the zinc and dilute sulphuric acid. Within the porous cup a strip of platinum dips into strong nitric acid. The hydrogen decomposes some of the nitric acid, taking oxygen from it to produce water and liberating red fumes of nitrogen tetroxide, which are unpleasant and hurtful. This battery gives very high electro-motive force.

FIG. 219.

Grove Cell.

BUNSEN'S BATTERY.—In this rods of carbon are substituted for the strips of platinum used in the Grove battery. Sometimes potassium bichromate

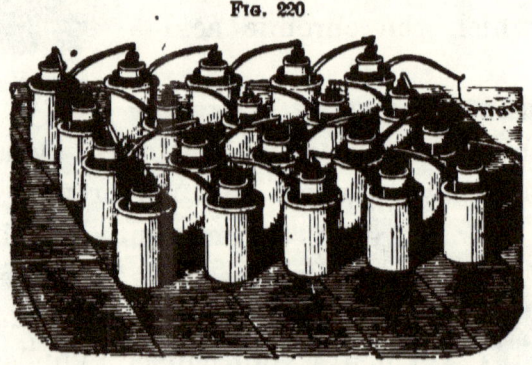

FIG. 220

Bunsen Battery.

solution is substituted for nitric acid in order to avoid the production of nitrous fumes. Fig. 220 shows a Bunsen battery arranged in series.

11. Effects of Voltaic Electricity.—(1.) PHYSICAL. —If a current of electricity is passed through a wire too small to conduct it readily, it is converted into heat. The poorer the conducting power of the wire, and hence the greater the resistance, the more marked the effect. With ten or twelve Grove's cups several inches of fine steel wire may be fused; and with a powerful battery, several yards of platinum wire may be made to glow with very brilliant effect, giving a steady light.*

In closing or breaking the circuit, we produce a spark, the size of which depends on the electromotive force and current strength of the battery. With several cells, beautiful scintillating sparks are obtained by fastening one pole to a file and rubbing the other upon it. When charcoal or gas-carbon electrodes are used with a powerful battery, on slightly separating the points, the intervening space is spanned by an arch of the most dazzling light (Fig. 221). The flame, reaching out from the positive pole like a tongue, vibrates around the negative pole, licking now on this side and now on that. The heat is intense. Platinum melts in it like wax

Fig. 221.

The Arc Light.

---

* Torpedoes and blasts are fired on this principle. Two copper wires leading from the battery to the spot are separated in the powder by a short piece of small steel wire. When the circuit is completed, the fine wire becomes red-hot and explodes the charge.

in the flame of a candle,* the metals burn with their characteristic colors; and lime, quartz, etc., are fused. The effect is not produced by burning the charcoal points, since in a vacuum it is equally brilliant.

(2.) CHEMICAL EFFECTS.—*Electrolysis* (to loosen by electricity) is the process of the decomposition of compound bodies by the voltaic current. If platinum electrodes be held a little distance apart in a cup of water mixed with sulphuric acid, tiny bubbles will immediately begin to rise to the surface. When the gases are collected, they are found to be oxygen and hydrogen, in the proportion of two volumes of the latter to one of the former.† In the electrolysis of

---

* To show the varying conducting power of the different metals, fasten together alternate lengths of silver and platinum wire and pass the current through them. The latter will glow, while the former, conveying the electricity more perfectly, will scarcely manifest its presence.

There are two forms of the electric light now used—the *arc* (shown in Fig. 221), where the current passes between two carbon points; and the *incandescent*, where the current heats to a dazzling white a carbon strip placed in the circuit. The former is employed in lighting streets, railroad stations, and large halls; the latter is generally used in dwellings, etc., as it gives a softer light, and is much more steady. Edison's Lamp consists of a tiny carbon loop placed in a glass globe from which the air has been so completely exhausted as to leave only $\frac{1}{1000000}$ of an atmosphere. When exposed to the air, the voltaic arc rapidly wastes the carbon points. Electric lamps have therefore been devised that, by a self-acting apparatus, keep the points at a proper distance from each other.

† If the copper poles be inserted, bubbles will pass off from the negative, but none from the positive pole, since the oxygen combines with the copper wire. That gas has no effect on platinum. The burning of an atom of zinc in the battery develops enough electricity to set free an atom of oxygen at the positive pole. It is interesting to notice that in the battery there is zinc burning, *i. e.*, combining with oxygen, but without light or heat; in the electric light the real force of the combustion is revealed. We may thus transfer the light and heat to a great distance from the place where they take their origin. The transmission of energy thus to a distance is better effected through electricity than through any other agency. Much ingenuity has been expended on machines for this purpose.

compounds, their elements are found to be in different electrical conditions. Hydrogen and most of the metals go to the negative pole, and are *electro-positive*. Oxygen, chlorine, sulphur, etc., go to the positive pole, and are therefore *electro-negative*.

On disconnecting the electrodes of the voltameter (Fig. 222) from the battery and joining them with a conductor, a current passes through the conductor

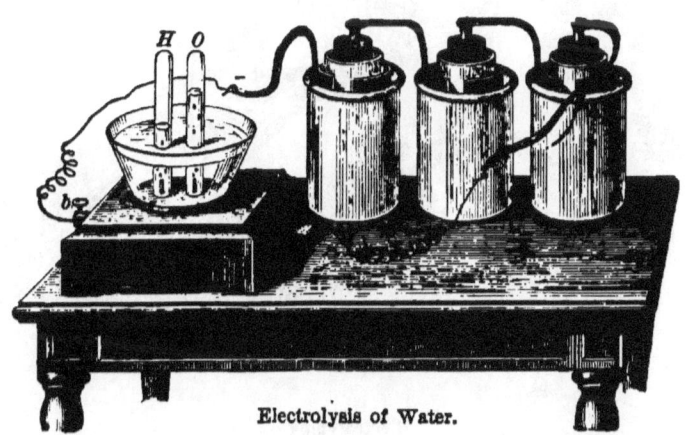

Electrolysis of Water.

from the electrode covered with oxygen to that covered with hydrogen. A voltameter thus charged constitutes a secondary cell in which the energy of a current may be stored up and again given out.*

Electrotyping is the process of depositing metals from their solutions by electricity. It is used in

---

* Faure's accumulator consists of two lead plates coated with red lead, rolled together with flannel between them, and immersed in dilute sulphuric acid. A current of electricity passed through such a cell changes a part of the red lead on the positive plate into peroxide of lead; and a part of the red lead on the negative plate into spongy metallic lead. A battery of these cells when freshly charged will retain its energy and produce a sustained current when desired.

copying medals, wood-cuts, types, etc. An impression of the object is taken with gutta-percha or wax. The surface to be copied is brushed with black-lead to render it a conductor. The mold is then suspended in a solution of copper sulphate, from the negative pole of the battery, and a plate

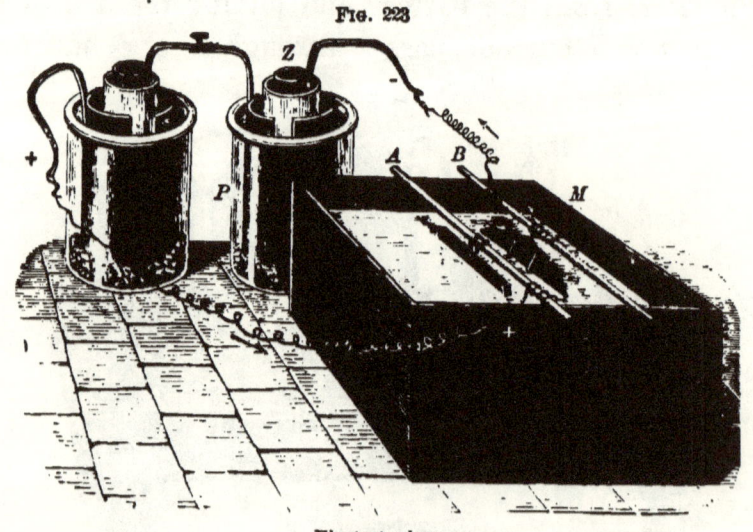

Fig. 223

Electrotyping.

of copper is hung opposite on the positive pole. The electric current decomposes the copper sulphate; the metal goes to the negative pole and is deposited upon the mold, while the acid, passing to the positive pole, dissolves the copper, and preserves the strength of the solution.*

* While the plate is hanging in the solution there is no noise heard or bubbling seen. The most delicate sense fails to detect any movement. Yet the mysterious electric force is continually drawing particles of *ruddy, solid* copper out of the *blue liquid*, and, noiselessly as the fall of snow-flakes, dropping them on the mold; producing a metal purer than any chemist can manufacture, spreading it with a uniformity no artist can attain, and copying every line with a fidelity that knows no mistake.

VOLTAIC ELECTRICITY. 325

Electro-plating is the process of coating with silver or gold by electricity. The metal is readily deposited on German silver, brass, copper, or nickel silver (a mixture of copper, zinc, and nickel). The objects to be plated are thoroughly cleansed, and then hung from the negative pole in a solution of silver, while a plate of silver is suspended on the positive pole. In five minutes a "blush" of the metal will be deposited, which conceals the other metal and is susceptible of polish.*

(3.) PHYSIOLOGICAL EFFECTS.—With a single cell no special sensation is experienced when the two poles are held in the hands. With a large battery a sud-

---

* Place in a large test-tube a silver coin with a little nitric acid. If the fumes of the decomposed acid do not soon rise, warm the liquid. When the silver is dissolved, fill the tube nearly full of soft water. Next drop hydrochloric acid into the liquid until the white precipitate (silver chloride) ceases to fall. When the chloride has settled, pour off the colored water which floats on top. Fill the tube again with soft water; shake it thoroughly; let it settle, and then pour off as before. Continue this process until the liquid loses all color. Finally, fill with water and heat moderately, adding potassium cyanide (the pupil will remember that this substance is exceedingly poisonous) in small bits as it dissolves, until the chloride is nearly taken up. The liquid is then ready for electro-plating. Thoroughly cleanse a brass key, hang it from the negative pole of a small battery, and suspend a silver coin from the positive pole. Place these in the silver solution, very near and facing each other. When well whitened by the deposit of silver, remove the key and polish it with chalk. In the arts the polishing is performed by rubbing with "burnishers." These are made of polished steel, and fit the surfaces of the various articles upon which they are to be used. It is said that an ounce of silver can be spread over two acres of surface. A well-plated spoon receives about as much silver as there is in a ten-cent piece. The only method of deciding accurately the amount deposited is by weighing the article before and after it is plated.—A vessel may be "gold-lined" by filling it with a solution of gold, suspending in it a slip of gold from the positive pole of the battery, and then attaching the negative pole to the vessel. The current passing through the liquid causes it to bubble like soda-water, and in a few moments deposits a thin film of gold over the entire surface.

den twinge is felt, and the shock becomes painful and even dangerous, especially if the palms are moistened with salt or acid water to increase the conduction. Rabbits which had been suffocated for half an hour, have been restored by an application of a strong voltaic current.

III. TRANSFORMATIONS OF ELECTRIC ENERGY.

1. Effect of a Voltaic Current on a Magnetic Needle.—If a wire conducting an electric current be placed over a poised magnetic needle, this tends to place itself at right angles to the wire.

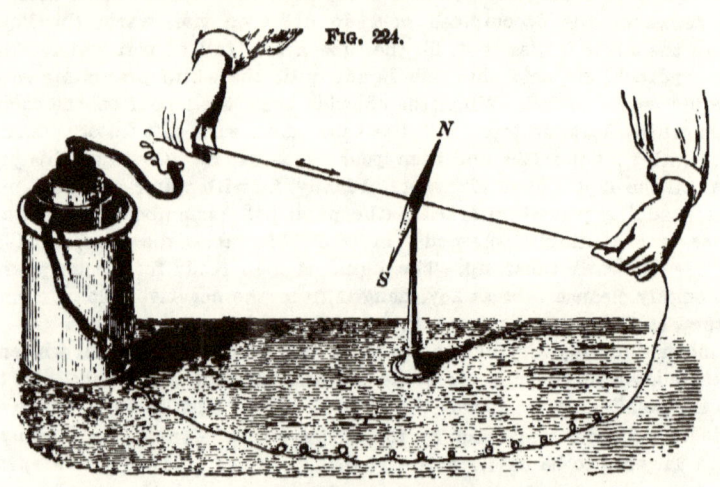

FIG. 224.

Effect of Current on Needle.

Assuming the direction of the current to be northward, the north pole of the needle will be turned toward the left. The same effect will be produced if the current pass southward under the needle, or vertically downward on the north side of it, or vertically upward on the south side of it. By reversing these conditions, the north pole of the needle will be turned toward the right. The play of the needle becomes thus a test of the presence and direction of an electric current.* The delicacy of

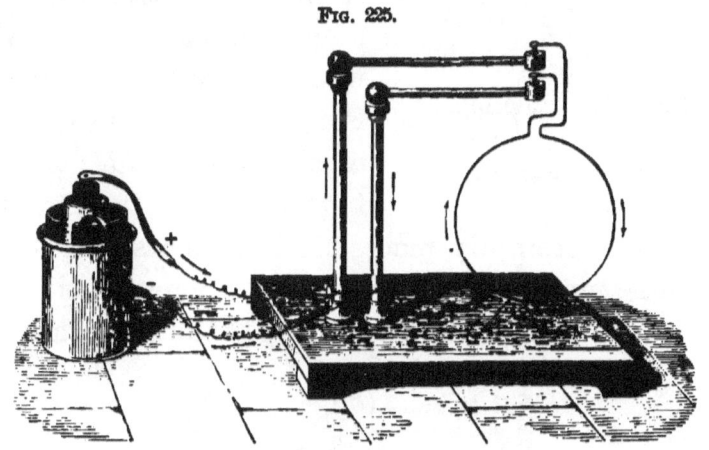

FIG. 225.

Pivoted Hoop of Conducting Wire.

this test is greatly increased if the wire, properly insulated, be coiled into a ring with many turns, at the center of which the needle is pivoted or suspended.

2. A Wire bearing a Current acts like a Magnet.—Let a copper-wire hoop be pivoted, as shown in Fig. 225, so that its plane is in a north and south

---

* Ampère gave a very convenient rule for determining the direction of the current from the motion of the magnetic needle. Imagine the current to be like a stream of water, with a little swimmer in it, facing the needle and swimming along with the current. The north pole of the needle will always turn *toward his left*. The pupil should try the experiment and test it in all possible ways.

direction. On passing an electric current through the hoop it slowly turns until its plane assumes an east and west position and its axis points north and south. Evidently the current-circle behaves like a magnet; one side of it being north, the other south-magnetic. A more effective arrangement consists of a wire wound spirally as shown in Fig. 226. A wire so wound is called a *solenoid*. On passing a current through the solenoid it promptly places itself with its axis in the magnetic meridian. With a pair of such solenoids (see Fig. 227) the experiments relating to the attraction and

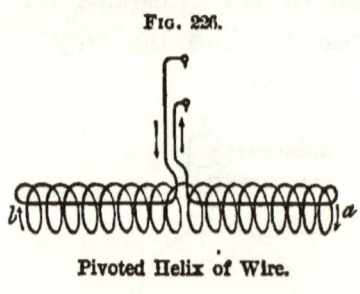

Fig. 226.

Pivoted Helix of Wire.

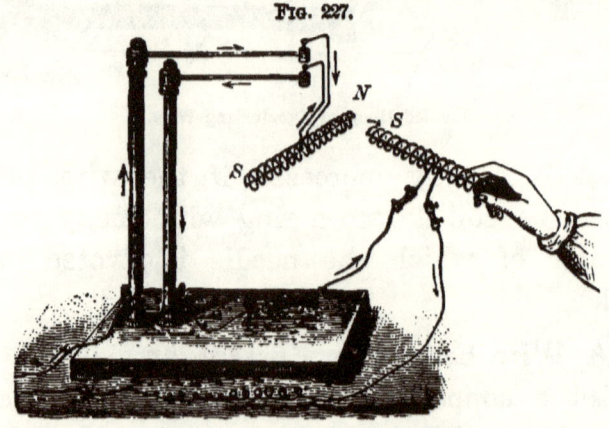

Fig. 227.

Two Helices acting like Magnets.

repulsion of poles of magnets may be repeated. The polarity of the solenoid depends upon the direction in which the current passes through it, and may be determined by the following rule: If on looking along the axis of the solenoid the current runs through it in a direction *contrary* to the direction of the hand of a watch (viewed face up), the observer is looking at the north end of the solenoid.

Fig. 228.

The Tangent Galvanometer.

3. The Tangent Galvanometer.—Any instrument designed to measure the length of an electric current is called a galvanometer. Of the many varieties, the tangent galvanometer is the most important. It consists of one or more coils of insulated wire wound upon a wooden hoop, at the center of which a small magnetic needle is pivoted or suspended (Fig. 228). The plane of the coils is made to coincide with that of the magnetic meridian. The earth's magnetism tends to keep the needle in this plane. The magnetic effect of a current tends to make it assume a position across this plane. Obeying both forces it assumes an oblique position, so as to make a measurable angle with the meridian. The strength of current is proportional to the tangent of this angle.

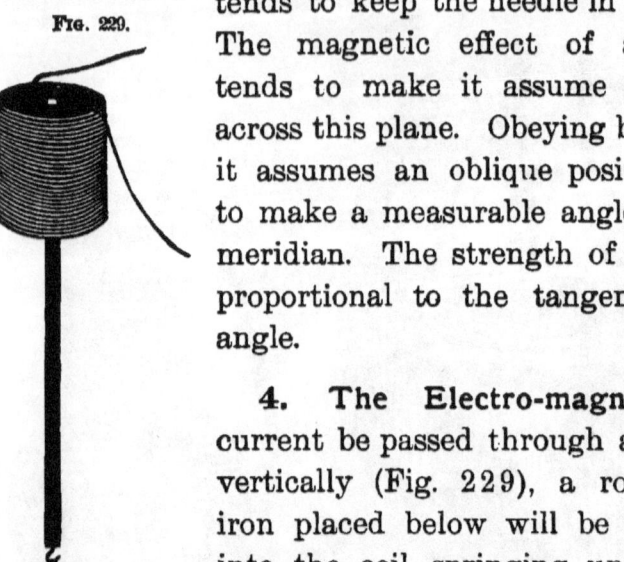

Fig. 229.

A Magnetic Mahomet's Coffin.

4. **The Electro-magnet.**—If a current be passed through a coil held vertically (Fig. 229), a rod of soft iron placed below will be drawn up into the coil, springing up as if endowed with life at the moment the current begins. It drops as soon as the circuit is broken.* Let a pair of such coils be fixed around the arms of a U-shaped rod of soft iron. This becomes a strong horseshoe magnet, whose strength comes and goes as the current is made or broken. It is therefore called an electro-magnet. Such magnets have been made strong enough to sustain a weight of several tons attached to the armature below.

5. **The Electro-magnetic Telegraph** depends on

---

* Thus is realized in science the fabulous story of Mahomet's coffin, which is said to have been suspended in mid-air.

the principle of closing and breaking the circuit at one station, and thereby making and unmaking an electro-magnet at the station with which communication is held. A single wire connects the two stations and is joined at each station to a key, a register (sounder), and a battery. One pole of each battery is connected with the ground. When a current is sent along the wire the circuit is completed through the earth. The key is used for sending messages; the register for receiving them.

Fig. 230.

The Electro-magnet.

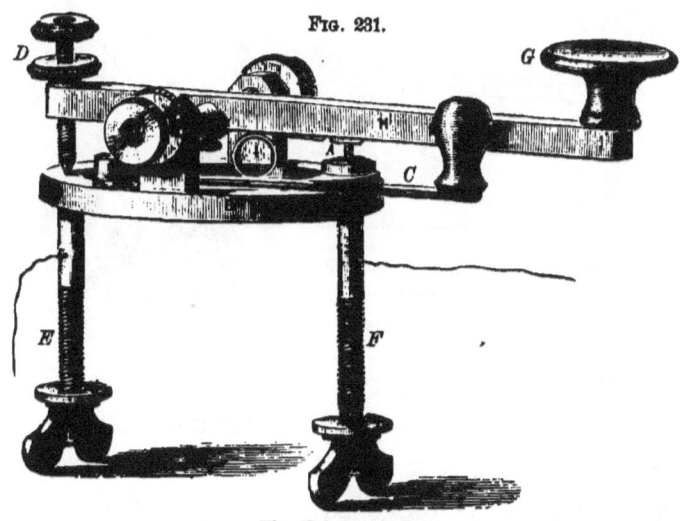

Fig. 231.

The Telegraph Key.

The key is shown in Fig. 231. E and F are screws which fasten the instrument to the table, and also hold the two ends of the wire. F is insulated by a ring of vulcanite where it passes through the table and the metal plate *B*. H is a lever with a finger-button *G*, a spring *I*, to keep it lifted, and a screw *D*, to regulate the distance it can move. At *A* is a break between two platinum points, which form the real ends of the wires. When *G* is depressed, the circuit is complete, and when lifted, it is broken. *C* is a *circuit-closer* that is used when the key is not in operation; the arm being pushed under *A* touches the platinum wire, and so completes the circuit. It is pushed out whenever the operator manipulates *G*. Then, by moving *G*, he can "close" or "open" the circuit at pleasure. He thus sends a message.

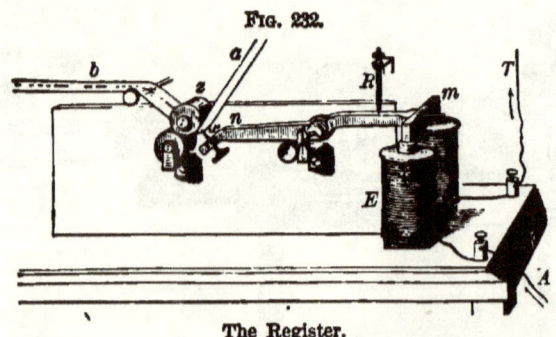

Fig. 232.

The Register.

The *register* contains an electro-magnet, *E* (Fig. 232). When the circuit is complete, the current, passing through the coils of wire at *E*, attracts the armature *m*. This elevates *n*, the other end of the lever *mn*, and forces the rounded point *x* firmly against the soft paper *a*. As soon as the circuit is broken, *E* ceases to be a magnet, and the spring *R* lifts the armature, drawing the point from the paper.

Clock-work attached to the rollers at z moves the paper along uniformly beneath the point x. When the circuit is completed and broken again instantly, there is a short dot made on the paper. This is called e; two dots, i; three dots, s; four dots, h. If the current is closed for a longer time, the mark becomes a dash, t; two dashes, m; a dot and a dash, a.

TABLE OF MORSE'S SIGNS.

a · —	j — · —	s · · ·
b — · · ·	k — · —	t —
c · · · ·	l — —	u · · —
d — · ·	m — —	v · · · —
e ·	n — ·	w · — —
f · — ·	o · · ·	x · — · ·
g — — ·	p · · · · ·	y · · · ·
h · · · ·	q · · — ·	z · · · ·
i · ·	r · · · ·	& · · · ·

A skillful operator becomes so accoustomed to the sound that the clicking of the armature is perfectly intelligible. He uses, therefore, simply a "*sounder*," *i.e.*, a register without the paper and clock-work attachment. Indeed, the register has now gone almost entirely out of use, and every operator is required to read by sound.

RELAY.—When the stations are more than fifty or sixty miles apart, the current becomes generally too weak to work the register. By substituting the *relay* for it in the line circuit, the force of a local battery may be employed to work the sounder or register.

In Fig. 233, which represents a relay, D is connected with the line wire, and C with the ground wire; A is connected with the positive pole of the local battery, and B with the register or sounder, and thence with the negative pole of this battery. The main current passes in at D, traverses the fine wire of the electro-magnet, K, and thence passes out at C to the ground. The armature E, playing to and fro as

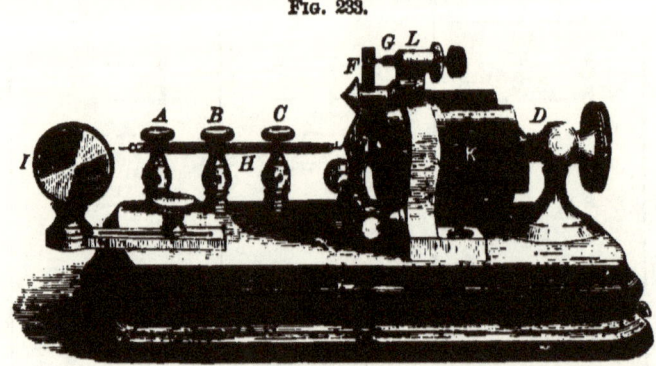

Fig. 233.

The Relay.

the current from the distant station passes through or is cut off, moves the lever F. This works on an axis at the lower end and is drawn back by the spring H, which is regulated by the thumb-screw I. As E is attracted, the circuit at G is closed; the current from A traverses a wire underneath, up F, and down L, and back through another wire underneath to B; thence to the electro-magnet of the sounder, which therefore attracts its armature.

The operator who sends the message simply completes and breaks the circuit with the *key;* the *armature* of the *relay*, at the station where the message is received, vibrates in unison with these movements; the *register* or *sounder* repeats them with greater force; and the second operator interprets their meaning.*

6. Ocean Cables.—The Atlantic and Indian Oceans have been spanned with cables of insulated wire for the transmission of telegraphic messages. The cable must be well insulated and very strong. In the middle is a bundle of copper wires, O (Fig. 234); this is buried within a sheathing of gutta-percha, C; around this is a group of cords of tarred hemp, H, to protect the gutta-percha; and, to give still further protection and strength, fifteen or twenty iron wires are twisted around the whole, so as to make it a rope about an inch thick.

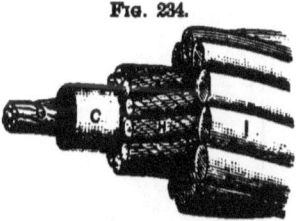

Fig. 234.

A Piece of Ocean Cable.

In signaling over a long cable, allowance has to be made for the great resistance of the long wire, and the slowness with which the circuit has to be operated. Instead of using a relay or sounder, it is necessary to use a delicate galvanometer as a receiver. A beam of light is reflected from a little

---

* The simple telegraph instrument is but one of a multitude of applications that have been made of the electro-magnet. By various ingenious devices it has become possible to send two, or even four, messages with reasonable rapidity over the same line at the same time. By one system, devised by Mr. Delany, as many as seventy-two circuits have been operated with a single instrument at the rate of two or three words per minute. The message is often printed at the moment it is received. From the Stock Exchange in New York hundreds of printed reports are thus sent at the same time to offices in various parts of the city.

mirror attached to the magnetic needle, and the swinging of the bright spot on a screen is interpreted as an alphabet. Or, a fine glass siphon tube is attached to a movable galvanometer coil which swings between the poles of an electro-magnet. Its short arm dips into a vessel of ink which is insulated and can be electrified. The long arm has its end over a strip of paper moved by clock-work. The electrifica-

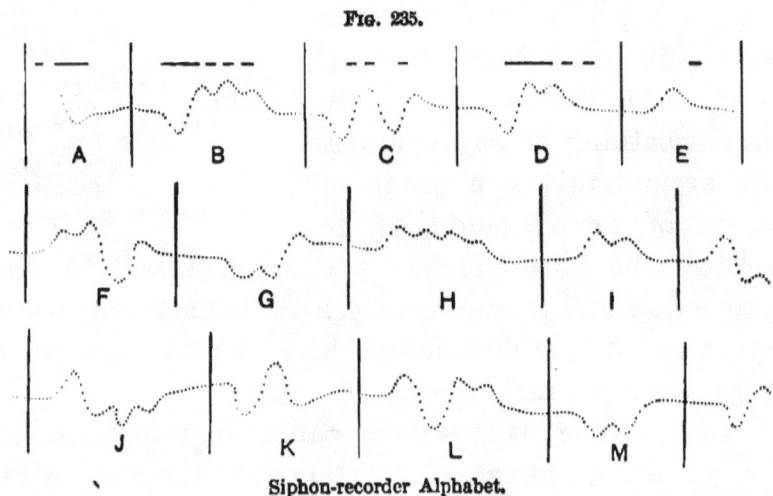

Fig. 235.

Siphon-recorder Alphabet.

tion of the ink causes it to issue in fine drops over the moving paper, and a sinuous line is recorded. This "Siphon-recorder" alphabet is partly shown in Fig. 235.

7. Electro-Magnetic Induction.— An electric current or a magnet produces in the space about it what is called a *field of force*. The greater the strength of the current or magnet, the greater is the force in the field. The farther we proceed from the current or magnet, the less becomes the force. A closed circuit in which no battery is included, placed in such a field will have a current *induced* in it whenever the force in the field acting on the circuit *varies* in strength. Whenever, therefore, the *inducing* current is made or broken, or the *inducing* magnet made or unmade, or when their strength is simply varied. Relative motion of the inducing current or magnet and the closed circuit will

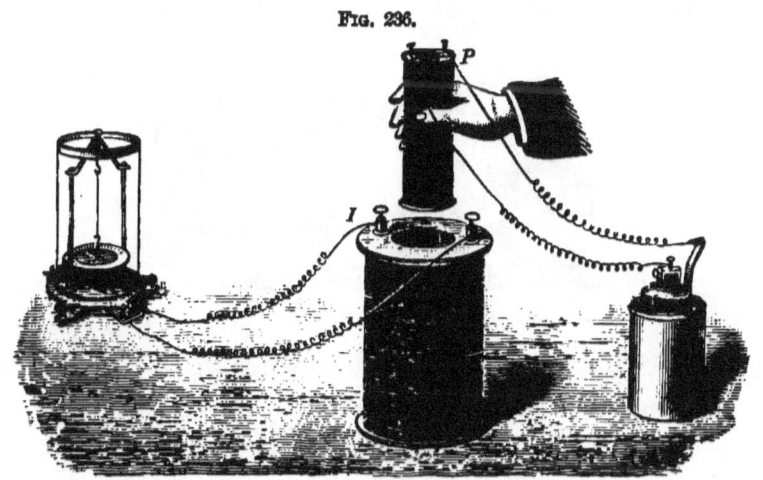

Fig. 236.

Current Induction.

also vary the force acting on the closed circuit and produce in it an induced current. At *every* such change a *momentary* current is induced in the closed circuit. Let the coil P (Fig. 236), which forms a closed circuit with the battery, be thrust into the coil I, which forms a closed circuit with the galvanometer. Instantly the needle turns, and then comes back to rest. Suddenly withdraw P. The needle turns in the opposite direction, and again comes back to rest. P is called the primary coil; I, the secondary or induction coil, because currents in it are *induced* by the motion of *P*. The current in *I* is opposite in direction to that of *P* when this is thrust in, and in the same direction when *P* is withdrawn. Similar effects are obtained if a magnet be substituted for *P*, or if in the first experiment the current be alternately made and broken in *P*.

8. **Ruhmkorff's Induction Coil** is provided with an automatic circuit-breaker, consisting of an electro-

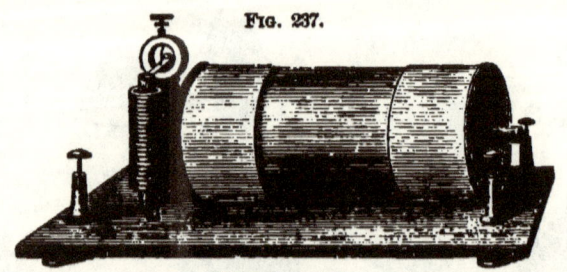

Fig. 237.

Induction Coil.

magnet whose current passes through a spring to which the armature is attached. When there is no current this spring touches a "contact point" which forms part of a circuit. On dipping the zinc into the acid of the battery, the current excites the electro-magnet. This attracts the armature, and thus removes the spring from the contact point. The circuit is hence broken, and the spring draws back the armature, making contact again. By this device, the current in the primary is automatically made and broken with great rapidity, and currents alternately in opposite directions are induced in the secondary. The intensity of these currents may be greatly increased by placing a bundle of soft iron wires in the axis of the primary. The inductive effect due to the magnetizing and demagnetizing of the iron wires by the making and breaking of the current in the primary is thus added to the inductive effect of the latter. The insulated wire of the secondary coil is long and fine, sometimes a hundred miles or more in length. The electro-motive force of the secondary current is enormously greater than that of the primary.* Connected with the poles of the battery is a condenser, which still further heightens the effect of the coil.

The induction coil is used for many purposes requiring high electro-motive force, and is usually more reliable than any machine generating electricity by friction. Beautiful effects are obtained by passing sparks from it through Geissler tubes. These are made of glass, and contain rarefied gases or vapors. The spark when passing through rarefied hydrogen assumes a brilliant red tint; through nitrogen, a gorgeous purple. With the proper degree of rarefaction it becomes stratified into bands across the tube.

9. The Telephone† is an instrument for utilizing magneto-electric currents and reproducing speech by their aid. Within a handle of vulcanite is a per-

---

* The largest induction coil ever constructed was made for Mr. Spottiswoode, an English physicist. Its secondary coil contained 280 miles of wire, wound in 340,000 turns, and its resistance exceeded 100,000 ohms. When worked with a Grove battery of 30 cells, it gave a spark 42 inches long, or considerably more than a yard in length. Coils containing 50 miles of wire are not uncommon; they yield sparks a foot or more in length. A Leyden jar interposed in the secondary circuit of such a coil is charged and discharged so rapidly as to make almost a continuous sound.

The action of the condenser is not easily explained in a few words to the elementary student. Consult Thompson's "Lessons in Electricity and Magnetism," pp. 363–365.

† A telephone in parts, ready to be put together by the experimenter is sold by the apparatus dealers. A simple but effective instrument can be

manent magnet (Fig. 238), around one pole, *N*, of which is an insulated coil, *C*, connected with the binding posts at the other end. A thin disk of soft iron, *BB*, is fixed across near the encircled magnet pole, and a mouth-piece, *A*, serves to direct the sound of the voice against the disk, which is thus made to vibrate. Disturbances are produced in the strength of the magnet, and corresponding currents traverse the wire. Passing through the coil of the distant telephone they vary the strength of its magnet. Minute clicking sounds are produced as the molecules of the magnet yield to these disturbances. The disk re-enforces these like a sounding-board, and gives out vibrations to the air, with such rapidity as to constitute a faithful reproduction of what was talked into the transmitting telephone.*

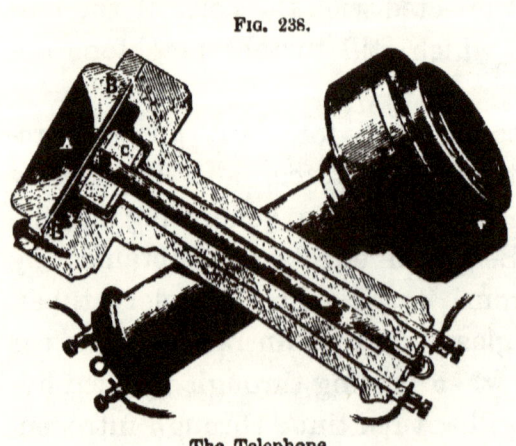

Fig. 238.

The Telephone.

made, at a slight expense, by a pupil with ordinary mechanical ability. One process, with illustrative drawings, is given in the "Popular Science Monthly," March, 1878, and another in the "Scientific American," Vol. XXXIX., No. 5. In Vol. XXXIX., No. 16, is also described a method of constructing a microphone; and in the "Scientific American Supplement," No. 133, is an account of a home-made phonograph. These numbers can be procured by any newsdealer. In using the telephone, two instruments exactly alike are employed. One is held to the mouth of the speaker, and the other to the ear of the listener.

* It should be observed that the presence of a disk is not necessary for the perception of sound from the receiving telephone. The motion is prob-

10. The Microphone is a modification of the telephone transmitter. It consists of a rod of gas carbon whose ends rest loosely in cups hollowed out of the same material, and these in turn fixed upon a sounding-board. The current from a battery passes through the microphone carbons and through a receiving telephone. If the sounding-board be made to vibrate in the least, whether by sound-waves or by slight mechanical motion, variations are produced in the pressure of the rod against its cups. Two pieces of carbon firmly in contact conduct electricity moderately well; but if the pressure between them is diminished, the resistance is increased and the current becomes fainter. The microphone is thus the last refinement of the telegraph combined with the telephone receiver. The sound of the voice, the patter of a fly's foot in walking over the sounding-board, or the gentlest ticking of a watch rested upon it, are thus made audible in a telephone many miles away.

Fig. 239.

Microphone.

ably among the molecules of the steel magnet, and conducted from them to the disk if this be added.

The telephone described in the text is the simplest that can be made. Many improvements have been effected in the instrument. The transmitting telephone is now generally made in such manner as to send an induced current, like that of the Ruhmkorff coil, through the line wire to the receiver.

11. The Magneto-electric Machine.—The production of a current in a closed circuit by moving it in the field of a magnet (p. 336, § 7) is realized in the construction of the magneto-electric machine. This machine consists of a powerful horseshoe magnet in front of which a pair of coils is made to rotate (Fig. 240). This pair is called the armature. Each coil contains a core of soft iron, which acquires and then

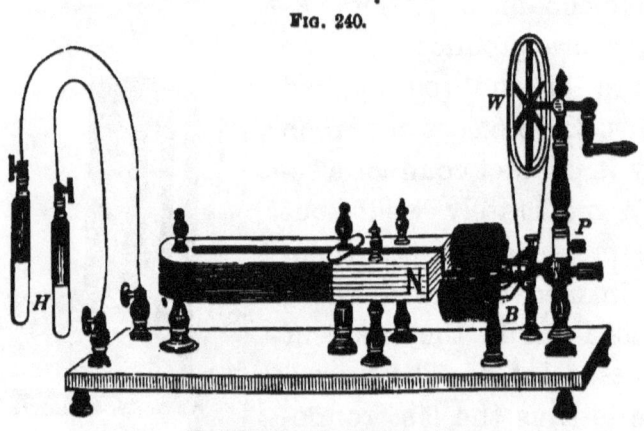

Fig. 240.

The Magneto-electric Machine.

loses magnetism, as it approaches, passes, and then recedes from a pole of the permanent magnet. In the coil these rapid variations of magnetic strength produce alternating currents whose electro-motive force is determined by the speed of rotation and the strength of the magnetic field. The two ends of the coil are connected with insulated plates of metal on opposite sides of the axle. On each of these a conducting spring presses, which carries the currents to the handles, H. This arrangement, called a *commutator*, is so adapted as to secure but one direction to the currents in the main wires. On taking hold of the handles while the shaft is rotated rapidly, a series of convulsive shocks is experienced.*

12. The Dynamo-electric Machine.—For the generation of currents to be employed in electric lighting it is necessary that they shall be continuous rather than intermittent. The name *dynamo* is applied to a development of the magneto-electric machine that accomplishes this result. The armature coils, in one type of these machines, are wound lengthwise upon a drum or cylinder, Fig. 241, which is revolved between the poles of a powerful electro-magnet called the field-magnet. On this drum a large number of coils may be wound, each with its own pair of commutator plates, these being so close together that the interval between two successive

---

* The machine represented in Fig. 240 is known as Clarke's machine, and was one of the first of its kind invented. Many improvements have been subsequently made. In 1866, Mr. Wilde discovered that if the induced current be passed through the coil of an electro-magnet, the strength it produces in this is far greater than that of the permanent magnet employed. An additional and larger armature was made to rotate in front of this electro-magnet, and the current induced in it was made to excite a second and still larger electro-magnet, whose armature then generated currents greatly stronger than any previously known. Such a machine, driven by a steam-engine of 15-horse power, produces an electric light dazzling as the noonday sun, throwing the flame of the street-lamps into shade at a quarter-mile distance. Its heat is sufficient to fuse a ¼-inch bar of iron fifteen inches long or 7 feet of No. 6 iron wire.—" A Yankee once threw the industrial world of Europe into a wonderful excitement by announcing a new theory of perpetual motion based on the magneto-electric machine. He proposed to decompose water by the current of electricity; then burn the hydrogen and oxygen thus obtained. In this way he would drive a small steam-engine, which, in turn, would keep the magneto-electric machine in motion. This would certainly be a splendid discovery. It would be a steam-engine which would prepare its own fuel, and, in addition, dispense light and heat to all around."—HELMHOLTZ.

currents is imperceptible. In Fig. 242, the end of the cylinder and of the group of commutator plates are seen between the large pole-pieces, N and S, of the field-magnets. The current is conducted off by the springs or "brushes," and passes through the coils of the field-magnet before reaching the main-line-wire. The pole-pieces never quite lose their magnetism, even after the machine is at rest. The energy of the induced current is at first wholly absorbed in exciting the field-magnet. This action, even though almost infinitely weak at first, increases until the magnet is as strong as possible; after which the energy is expended in doing work on the main circuit.*

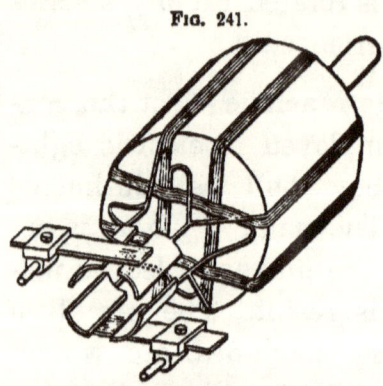

Fig. 241.

Drum Armature and Four-part Commutator.

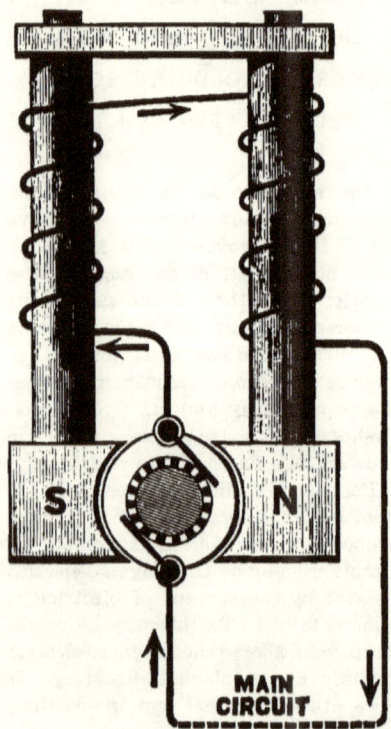

Fig. 242.

Diagram of a Series Dynamo.

* In the frontispiece is a picture of the Weston Dynamo, such as is used for producing the electric lights on the great bridge between New York and Brooklyn. Each pole-piece is attached to two coils which are so wound that both have the same effect on it. The end of

13. Electric Motors.—The action of a dynamo is reversible. By revolving its armature we obtain a current, and by sending a current through its armature we cause it to revolve. The same machine may, therefore be used either as a dynamo or as a motor. The dynamo converts the mechanical energy of the source of power into the energy of the electric current; the motor converts the energy of the electric current into mechanical energy. On this principle depends the *transmission of power*.

14. The Electric Light.—According to the method adopted in winding the armature, a dynamo-machine may give currents of high or of low electro-motive force. Two corresponding kinds of lamp are used. To produce the *arc light*, a current of high electro-motive force is sent through a pair of carbon rods, which are then drawn slightly apart. Particles of carbon are made white-hot, and even turned into vapor, which is thrown from one rod to the other in the direction of the current. The path of the glowing vapor is curved, and hence this is called the voltaic arc. It is the most brilliant artificial light known, but unsteady because the arc leaps from side to side as the carbons become wasted away. An automatic regulator is employed to keep them at the proper distance apart. The arc light is excellent for lighting streets, halls, and other large public places. For domestic use, the *incandescence* lamp is better.

the drum armature is covered with radiating conductors, which connect the coils with the commutator plates. One of the brushes is seen pressing on these plates, and is connected with the insulated wire that conducts the current away.

In this a current of low electro-motive force passes through a filament of carbon inclosed in a globe from which most of the air has been withdrawn. The filament glows with a soft and steady light, which is much inferior in brilliancy to the arc light. Although only $\frac{1}{1000000}$ of the air remains in the globe, the filament is slowly burned away, and has to be replaced with a new one.

15. Thermo-electricity.—In the electric lamp the energy of a current is changed into heat and light. Conversely, heat and light (radiant energy) may be changed into electric energy. If the end of an iron wire be connected with that of a copper or German-silver wire, the other ends being attached to a galvanometer, the needle will swing aside when the joined ends are heated. This effect is increased if the junction be made with bismuth and antimony. A current flows at the junction from bismuth to antimony, thence through the galvanometer back to the bismuth.

A THERMO-ELECTRIC PILE consists of alternate bars of antimony and bismuth soldered together, as shown in Fig. 243. When mounted for use, the couples are insulated from each other and inclosed in a copper frame, P. If both faces of the pile are equally heated, there is no current. The least variation of temperature, however, between the two is indicated by the

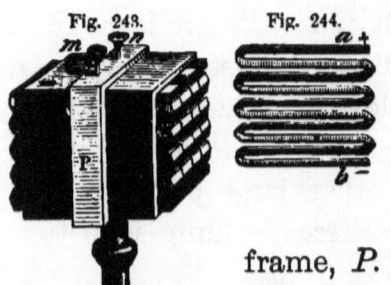

Fig. 243. Fig. 244.

Thermopile.

TRANSFORMATIONS OF ENERGY. 347

flow of electricity. Wires from a, the positive pole, and b, the negative, connect the pile with the galvanometer. This furnishes a test of change of temperature. A fly walking over the face of the pile by its warmth will move the needle, if the galvanometer be very delicate. When skillfully used, the thermopile serves as a very sensitive thermometer.

THE BOLOMETER is an instrument devised for the detection of very faint variations of temperature. A platinum or iron wire opposes much more resistance to the passage of an electric current when hot than when cold. The current is made to divide between two conductors. These are connected by a crosswire, or "bridge," with a galvanometer interposed. If the current in the two branches be equal, the galvanometer is not affected; but, if unequal, a crosscurrent deflects the galvanometer needle. By heating one branch slightly the balance is disturbed, and the difference of temperature is read in the deflection of the needle.*

16. Animal Electricity.—The human body is often electrified. Many animals, especially when angry or otherwise excited, give evidence of being electrified.

* This instrument was invented by Professor Langley at the Alleghany Observatory near Pittsburg. It was used in examining the invisible parts of the solar spectrum, where lines and bands were discovered whose presence could not be detected with the most delicate thermopile. The invisible part of the spectrum was thus found to be much more extensive than the visible part, while the most intense heat as well as light is found in the region colored greenish-yellow. The bolometer is capable of revealing a change of temperature of $.00001°$ C. Professor Langley has discovered by this means that the highest temperature of the moon scarcely, if at all, exceeds that of the human body, and that the temperature of outer space is nearly as low as the absolute zero of temperature, $-273°$ C.

Certain fish have the property of giving, when touched, a shock like that from a Leyden jar. The torpedo and the electrical eel are most noted. The former is a native of the Mediterranean, and its shock was anciently prized as a cure for various diseases. The latter is abundant in certain South American waters. A specimen of this fish, forty inches in length, was estimated by Faraday to emit a spark equal to the discharge of a battery of fifteen Leyden jars of 3,500 square inches surface.

SUMMARY.

ELECTRICITY is a form of energy that may be manifested as an accompaniment of friction, of chemical action, of the motion of magnets, of variations in temperature, or of animal excitement. It exhibits a certain kind of duality in its effects, and hence the names positive and negative electricity are used to express the contrast. Many considerations point to the conclusion that the molecules of a charged body are in a condition of strain. This condition can be communicated by induction through a "dielectric," which itself becomes strained while thus acting as a medium. By taking advantage of a proper dielectric, such as glass, electrical energy may be stored up for subsequent use, as in the Leyden jar.

Voltaic electricity has its origin in chemical action, or in contact of different metals, or in both. The essentials of an ordinary battery for its development are two substances, which are unequally affected by a chemical agent. One of these is at higher potential than the other, and neutralization is effected by the passage of a current, continually renewed, from the body at high potential, through the best conductor, to the body at low potential. This difference of potential, however, is very slight in comparison with that developed by friction and induction, as in the Holtz or Voss machine. Voltaic electricity is more manageable, more reliable, more convenient, more generally available than frictional electricity. Electricity may be transformed, under appropriate conditions, into mechanical motion, magnetism, sound, heat, or light. Among its most important applications to the purposes of practical life are the telegraph, the electrotype, the telephone, distribution of power, and the electric light.

HISTORICAL SKETCH.

THALES (6th cent. B. C.), one of the seven wise men, knew that when amber is rubbed with silk it will attract light bodies, as straw, leaves, etc. This property was considered so marvelous that amber was supposed to possess a soul. From the Greek name of the substance (elektron) our word electricity is derived. This simple phenomenon constituted all that was known until the 16th century, when William Gilbert, physician to Queen Elizabeth, made many valuable experiments. He discovered that amber was by no means the only substance which can exhibit electrical manifestations when rubbed, and he examined into the conditions favorable to electrical phenomena. Among the most important of these he found to be the dryness of the atmosphere. Francis Hawksbee called attention to the resemblance between the electric spark and lightning, and invented an electric machine, in which the hands were used as rubbers. Stephen Gray, in the 18th century, discovered the difference between conductors and non-conductors, that an electric charge is at the surface, and that the human body can be electrified. Dufay discovered that there are two manifestations of electricity, which he called vitreous and resinous, and considered them to be fluids. Kinnersley, the friend and associate of Franklin, recognized that these two electricities were nothing else than what Franklin had already called positive and negative charges. The Leyden jar was invented in 1745, probably by several persons about the same time; it was first exhibited and used in experiment by Muschenbroeck, at Leyden, in Holland. By the use of it students of electricity were able to gather the mysterious "virtue" or "effluvia" in much larger quantities, and to produce effects never imagined before, such as the firing of gunpowder. Experiments were made about this time to ascertain the rate of transmission of electricity from a Leyden jar through a metallic conductor. A wire more than two miles long was employed; through this the discharge appeared to be absolutely instantaneous.

In 1749, Benjamin Franklin wrote from Philadelphia to Peter Collinson at London, as follows:

"Chagrined a little that we have hitherto been able to produce nothing in this way of use to mankind, and the hot weather coming on, when electrical experiments are not so agreeable, it is proposed to put an end to them for this season, somewhat humorously, in a party of pleasure on the banks of the *Skuylkil*. Spirits, at the same time, are to be fired by a spark sent from side to side through the river, without any other conductor than the water; an experiment which we some time since performed, to the amazement of many. A turkey is to be killed for our dinner by the *electrical shock*, and roasted by the *electrical jack* before a fire kindled by the electrical bottle (Leyden jar); when the healths of all the famous electricians in England, Holland, France, and Germany are to be drank in *electrified bumpers*, under the discharge of guns from the *electrical battery*."

About 1752, Franklin proved the identity of lightning and frictional electricity by means of a kite made of a silk handkerchief and with a pointed wire at the top. He elevated this during a thunder-storm, tying at the end of the hemp string a key, and then insulating the whole by fastening it to a post with a long piece of silk lace. On presenting his knuckles to the key, he obtained a spark. He afterward charged a Leyden jar, and performed other electrical experiments in this way. These attempts were attended with very great danger. Prof. Richman, of St. Petersburg, drew in this manner from the clouds a ball of blue fire as large as a man's fist which struck him lifeless. Shortly after the famous experiments of Franklin, the Frenchman, Coulomb, established the law of electric attraction and repulsion, showing that it was the same as that of gravitation, light, and heat, the law of inverse squares.

In the year 1790, Galvani was engaged in some experiments on animal electricity. For this purpose he used frogs' legs as electroscopes. He had hung several of these upon *copper* hooks from the *iron* railing of the balcony, in order to see what effect the atmospheric electricity might have upon them. He noticed, to his surprise, that when the wind blew them against the iron supports, the legs were convulsed as if in pain. After repeated experiments, Galvani concluded that this effect was produced by what he termed animal electricity, that this electricity is different from that caused by friction, and that he had discovered the agent by which the will controls the muscles. Volta rejected the idea of animal electricity, and held that the contact of dissimilar metals was the source of the electricity, while the frog was "only a moist conductor, and for that purpose was not as good as a wet rag." He applied this view to the construction of "Volta's pile," which is composed of plates of zinc and copper, between which are laid pieces of flannel moistened with an acid or a saline solution (Fig. 245). This theory is substantially the one held at the present time, though we now know that there must be chemical action to continue the supply.

FIG. 245.

Volta's Pile.

From the earliest times in which the knowledge of electricity began to be definite, impostors and half-educated people circulated marvelous stories about its value as a panacea for all kinds of disease. Many supposed that deafness and dimness of sight might be cured by the use of the electric spark. Franklin remarked of this, "it will be well if perfect blindness be not the consequence of the experiment." In the hands of experienced physicians electricity has been used with good effect, but to-day, as in Franklin's time, the name often serves as a cloak for ignorance or trickery.

Electricity and magnetism were studied as distinct branches until 1820, when Oersted of Copenhagen discovered the phenomenon shown in Fig. 224. This was published every-where, and excited the deepest interest of scientific men. In the fruitful mind of Ampère the experiment bore abundant fruit. He discovered that two parallel wires conveying an electric current in the same direction attract each other, and when in opposite directions, repel each other. From this he generalized the entire subject. Prof. Henry next exhibited the wonderful power of the electro-magnet, and invented the electro-magnetic engine. Scientific men in all parts of the world were now gathering the material necessary for the invention of the electric telegraph. It fell to Samuel F. B. Morse to make this knowledge practical, and in 1837 he exhibited in New York a working instrument. An experimental line between Washington and Baltimore was completed in 1844, and, on May 27th of that year, was sent the first message ever forwarded by a recording telegraph.

Consult Maxwell's "Electricity and Magnetism"; Tyndall's "Lessons in Electricity"; "Faraday's "Lectures on the Physical Forces" and "Researches in Electricity"; Noad's "Manual of Electricity"; Art. on the Microphone, in "Scribner's Monthly," Vol. XVI., p. 600; Prescott's "The Speaking Telephone, Talking Phonograph," etc.; Foster's "Electrical Measurements," in "Science Lectures at South Kensington," Vol. I., p. 264; Thomson's "Papers on Electrostatics and Magnetism"; Guillemin's "The Forces of Nature" and "The Applications of Physical Forces"; "American Cyclopedia," Articles on Electricity, Magnetism, Electro-magnetism, etc.; Smith's "Manual of Telegraphy"; Jones' "Historical Sketch of Electric Telegraph"; Watts' "Electro-metallurgy"; "Barnes' Hundred Years of American Independence," Sec. on Morse, p. 442; "Fourteen Weeks in Zoology," Sec. on Torpedo, p. 186; Gordon's "Electricity and Magnetism"; Hospitalier's "Modern Applications of Electricity" (specially commended for latest discoveries); Urbanitzki's "Electricity in the Service of Man"; Mendenhall's "A Century of Electricity"; Thompson's "Dynamo-electric Machinery"; Daniell's "Principles of Physics," and Anthony and Brackett's "Text-book of Physics."

CONCLUSION.

"Science is a psalm and a prayer."—PARKER.

NOWHERE in nature do we find chance. Every event is governed by fixed laws. If we would accomplish any result or perform any experiment, we must come into exact harmony with the universal system. If we deviate from the line of law by a hair's breadth, we fail. These laws have been in operation since the earliest beginnings of the development of our world, and all the discoveries of science prove them to extend to the most distant star in space. A child of to-day amuses itself with casting a stone into the brook and watching the widening curves; little children of ten thousand years ago may have done the same. A law of nature has no force of itself; it is but *the manner in which force acts*.

We can not create force. We find it every-where in Nature; so that matter is not dumb, but full of inherent energy. A tiny drop of dew sparkling on a spire of grass is instinct with power: Gravity draws it to the earth; Chemical Affinity binds together the atoms of hydrogen and oxygen; Cohesion holds the molecules of water, and gathers the drop into a globe; Heat keeps it in the liquid form; Adhesion causes it to cling to the leaf. If the water be decomposed, Electricity will be set free; and from this, Heat, Light, Magnetism, and Motion can be produced. Thus the commonest object becomes full of fascination to the scientific mind, since in it reside the mysterious forces of Nature.

These various forces can be classified either as attractive or repellent. Under their influence the atoms or molecules resemble little magnets with positive and negative poles. They approach or recede from one another, and so tend to arrange themselves according to some definite plan. "The atoms march in time, moving to the music of law." A crystal is but a specimen of "molecular architecture" built up by the forces with which matter is endowed. Forces continually ebb and flow, but the sum of energy through the universe remains the same. In time all the possible changes may be rung, and the various forms of energy subside into one uniformly-diffused heat-quiver, but in that will exist the representation of all the forces which now animate creation.

XI.
APPENDIX.

APPENDIX

APPENDIX.

QUESTIONS.

THE following questions are those which the author has used in his classes, both as a daily review and for examination. A standing question, which has followed every other question, has been: "*Can you illustrate this?*" Without, therefore, a particular request, the pupil has been accustomed to give as many practical examples as he could, whenever he has made any statement or given any definition.

I. Introduction. — Define matter. A body. A substance. Name and define the two kinds of properties which belong to each substance. State the suppositions of the Atomic Theory. What is a molecule? An atom?

Describe the two kinds of change to which matter may be subjected. What is the principal distinction between Physics and Chemistry? Mention some phenomena which belong to each. Why are these branches intimately related?

Name the general properties of matter. Define magnitude Size. Distinguish between size and mass. Why are feathers light and lead heavy? Why is it necessary to have a standard of measure? What are the French and English standards? Give the history of the English standard. Is the American yard an exact copy of the English? Give an account of the French system. By what name is this system commonly known? Is either of these systems founded on a natural standard? Why is it desirable to have such a standard?

Define Impenetrability. Give some apparent exceptions, and explain them. Define Divisibility. Is there any limit to the divisibility of matter? Define Porosity. Is the word *porous* used here in its common acceptation? What practical use is made in the arts of the property of porosity? Describe the experiment of the Florence academicians.

Define Inertia. Does a ball, when thrown, stop itself? Why is it difficult to start a heavy wagon? Why is it dangerous to jump from the cars when in motion? (Compare First Law of Motion.) Define Indestructibility. Has the earth at all times contained the same quantity of matter that it does now?

Name the specific properties of matter. Define Ductility. How is iron wire made? Platinum wire? Gilt wire? Define Malleability.

Describe the manufacture of gold-leaf. Is copper malleable? Define Tenacity. Name and define the three kinds of Elasticity. Illustrate the elasticity of compression as seen in solids. In liquids. In gases. What is said about the relative compressibility of liquids and gases?

Illustrate the elasticity of expansion as seen in solids, liquids, and gases. Define Elasticity of Torsion. What is a Torsion balance? Define Hardness. Does this property depend on density? Define Density. Define Brittleness. Is a hard body necessarily brittle? Name a brittle and a hard body.

II. **Motion and Force.**—Define motion, absolute and relative. Rest. Velocity. Force. What are the resistances to motion? Tell what you can about friction. Why does oil diminish friction? What uses has friction? What law governs the resistance of air or water? Define Momentum.

Show that motion is not imparted instantaneously. State the three laws of motion and the proof of each. If a ball be fired into the air when a horizontal wind is blowing, will it rise as high as if the air were still? Describe the experiments with the collision balls. Give practical illustrations of action and reaction. If a bird could live, could it fly in a vacuum? Define compound motion.

Define the so-called "parallelogram of forces." The resultant. How can the resultant of two or more forces be found? Name some practical illustrations of compound motion. What is the "resolution of forces"?

Show how one vessel can sail south and another north,

driven by the same westerly wind. Explain how a kite is raised. Explain the "split-shot" in croquet.

Explain the towing of a canal-boat. Describe how motion in a curve and circular motion are produced. Explain the centripetal and centrifugal forces.

Show when the centrifugal force becomes strong enough to overcome the force of Cohesion. Of Adhesion. Of Gravity. Apply the principle of circular motion to the revolution of the earth about the sun. What effect does the revolution of the earth on its axis have upon all bodies on the surface?

What would be the effect if the rotation were to cease? Describe the action of the centrifugal force on a hoop rapidly revolved on its axis. Define reflected motion. Give its law.

What is Energy, in the Physical sense of the word? To what is it proportional? Name and define the two forms of energy. How may one form be changed into the other?

What is the law of the Conservation of Energy? What did Faraday say with regard to this law?

III. Attraction. *1.* MOLECULAR FORCES.—Define a molecular force. What two opposing forces act between the molecules of matter? How is this shown? What is the repellent force? Name the attractive forces. Which of these belong to Physics?

1. *Cohesion.*—Define. What are the three states of matter? Define. How can a body be changed from one state to another?

Show that cohesion acts only at insensible distances. Explain the process of welding. Why do drops of dew, etc., take a globular form? Why do not all bodies have this form? Illustrate the tendency of matter to a crystalline structure.

Has each substance its own form? Why is not cast-iron crystalline? Why do cannon become brittle after long use?

Describe the process of tempering and annealing. Explain the Rupert's Drop. How is glassware annealed?

2. *Adhesion.*—Define. What is the theory of filtering through charcoal? Of what use is soap in making bubbles? Define Capillary Attraction. Why will water rise in a glass tube, while mercury will be depressed? Is a tube necessary to show capillary attraction? What is the law of the rise in tubes?

Give practical illustrations of capillary action. Why will not old cloth shrink as well as new, when washed? What is the cause of solution? Why is the process hastened by pulverizing?

Tell what you can about gases dissolving in water. Why does the gas escape from soda-water as soon as drawn? Why do pressure and cold favor the solution of a gas? Describe the diffusion of liquids. Of gases.

Describe the osmose of liquids. Of gases. Why do rose-balloons lose their buoyancy? What is the difference between the *osmose* and the *diffusion* of gases?

2. Gravitation.—How does Gravitation differ from Cohesion and Adhesion? What is the law of gravitation? Why does a stone fall to the ground? Will a plumb-line near a mountain hang perpendicularly? Why do the bubbles in a cup of tea gather on the side? How is the earth kept in its place? Define Gravitation. Gravity. Weight.

State the three laws of weight. What is a vertical or plumb-line?

Describe the "guinea-and-feather experiment." What does it prove? Describe Atwood's machine. Deduce the formulas for falling bodies.

How can the time of a falling body be used for determining the depth of a well? How does gravity act upon a body thrown upward? What velocity must be given to a ball to elevate it to any point? How high will it rise in a given time? When it falls, with what force will it strike the ground? Define the Center of Gravity. The line of direction. The three states of equilibrium.

How may the center of gravity be found? Give the general principles of the center of gravity. Describe the leaning tower of Pisa. State some physiological applications of the center of gravity. Why do fat people generally walk so erect?

Define the Pendulum. Arc. Amplitude. What are isochronous vibrations? State the three laws of the pendulum. Who discovered the first law? What is the center of oscillation? How is it found? What is the center of percussion?

Describe the pendulum of a clock. How is a clock regulated? Does it gain or lose time in winter? Describe the gridiron pendulum. The mercurial pendulum. Name the various uses of the pendulum. Describe Foucault's experiment.

IV. **The Elements of Machines.**—Name and define the elements of machinery. Do the "powers," so called, produce energy? What is the law of mechanics? Illustrate the law. What is a lever? Describe the three classes of levers. The law of equilibrium.

What is the advantage peculiar to each class? Describe the steelyard as a lever. What effect does it have to reverse the steelyard? Describe the arm as a lever. (See "Hygienic Physiology," p. 34.) Would a lever of the first class answer the purpose of the arm? Describe the compound lever.

Describe the hay scale. The wheel and axle. Its law of equilibrium. Describe a system of wheel-work. At which arm of the lever is the P applied?

Describe the various uses of the inclined plane. Its law of equilibrium. What velocity does a body acquire in rolling down an inclined plane? Give illustrations.

Describe the screw. Its uses. Its law of equilibrium. How may its power be increased? What limit is there? Describe the wedge. Its uses. Its law of equilibrium. How does it differ from that of the other powers? Describe the pulley. The use of fixed pulleys. Is there any gain of P in a fixed pulley?

What is the use of a movable pulley? Describe a movable pulley as a lever. Give the general law of equilibrium in a combination of pulleys. What are cumulative contrivances? Is perpetual motion possible? Why?

V. **Pressure of Liquids and Gases.** 1. HYDROSTATICS.—Define. What liquid is taken as the type? What is the first law of liquids? Explain. Illustrate the transmission of pressure by water. Show how water is used as a mechanical power. Describe the hydrostatic press. Give its law of equilibrium.

What are the uses of this press? What pressure is sustained by the lower part of a vessel of water, when acted on by gravity alone? How does this pressure act? State the four laws which depend on this principle, and illustrate them. What is the weight of a cubic foot of sea water? Fresh water? What is the pressure at two feet? Give illustrations of the pressure at great depths. Describe the hydrostatic bellows. Its law of equilibrium. What is the "hydrostatic paradox"? Give illustrations. Give the principle of fountains. How high will the water rise? How do modern engineers carry water across a river? Did the ancients understand this principle? Give the theory of the Artesian well, and of ordinary wells and springs.

Give the rule for finding the pressure on the bottom of a vessel. On the side. Define the water level. Is the surface of water horizontal? If it were, what part of an approaching ship would we see first? Describe the spirit-level. Define specific gravity. What is the standard for solids and liquids? For gases? Explain the buoyant force of liquids.

What is Archimedes' law? Describe the "cylinder-and-bucket experiment." What does it prove? Give the method of finding the specific gravity of a solid. A liquid.

Is it necessary to use a specific gravity flask holding just 1,000 oz., or would any size answer? Suppose the solid is lighter than water and will not sink, what can you do? Explain the hydrometer. How can you find the weight of a given volume of any substance? The volume of any given weight? The exact volume of a body? Illustrate the action of dense liquids on floating bodies. Why will an iron ship float on water? Where is the center of gravity in a floating body? How do fish sink at pleasure?

2. HYDRODYNAMICS.—Define. To what is the velocity of a jet equal? How is the velocity found? Give the rule for finding the quantity of water which can be discharged from a jet in a given time. What is the effect of tubes? Tell something of the flow of water in rivers.

Name and describe the different kinds of water-wheels. Which is the most valuable form? Describe Barker's Mill. How are waves produced? Explain the real motion of the water. How does the motion of the whole wave differ from that of each particle? How is the character of waves modified near the shore? What is the extreme height of "mountain waves"? Define like phases. Unlike phases. A wave-length.

What is the effect if two waves with like phases coincide? With unlike phases? What is this termed?

3. PNEUMATICS.—Define. What principles are common to liquids and gases? What gas is taken as the type? Describe the air-pump. Can a perfect vacuum be obtained in this way? What is the condenser? Its use? Prove that the air has weight.

Show its elasticity and compressibility. Describe the bottle-imps. What principles do they illustrate? Show the expansibility of the air.

Describe the experiments with the hand-glass. The principle of Hiero's fountain. The Magdeburg hemispheres. What do they prove? Show the upward pressure of the air.

The buoyant force of the air. Would a pound of feathers and a pound of lead balance, if placed in a vacuum? On what principle does a balloon rise? What is the amount of the pressure of the air? Describe the experiment illustrating this. Where do these figures apply?

Describe how the pressure of the air continually varies. Explain Mariotte's (called also Boyle's) law. Describe the barometer. Its uses. Are the terms "fair," "foul," etc., often placed on the scale, to be relied upon? Why is mercury used for filling the barometer? Describe Otto Guericke's barometer.

Describe the action of the lifting-pump. The force-pump. The fire-engine. Compare the action of the lifting-pump with that of the air-pump. What is the siphon? Explain its theory.

Describe the pneumatic inkstand. The hydraulic ram. The atomizer. Show how a current of air drags with it the still atmosphere. What opposing forces act on the air? How high does the air extend? How does its density vary?

VI. Acoustics.—Define. Name and define the two senses of this word. May not the terms "light," "heat," etc., be used in the same way? Illustrate the formation of sound by vibrations.

Show how the sound of a tuning-fork is conveyed through the air. The report of a gun. The sound of a bell. The human voice. Define a sound-wave. In which direction do the molecules of air vibrate? In what form do the waves spread? Can a sound be made in a vacuum? Can a sound come to the earth from the stars?

How do sounds change as we pass above or below the sea-level? Upon what does the velocity of sound depend? Why is this? At what rate does sound travel in the air? In water? In the metals? In iron? What effect does temperature have on the velocity of sound?

Do all sounds travel at the same rate? How does the velocity of sound enable us to determine distance? Upon what does the intensity of sound depend? At what rate does it diminish? Why?

Explain the speaking-tube. The ear-trumpet. Describe Biot's experiment in the water-pipes of Paris. The speaking-trumpet. What is the refraction of sound?

Define reflection of sound. What is the law? Give some curious instances of reflection. What is the shape of a whispering-gallery? Illustrate the decrease of sound by repeated reflection. Why are sounds more distinct at night than by day? Is it desirable to have a door or a window behind a speaker? What causes the "ringing" of a sea-shell? How are echoes produced? When is the echo repeated? Illustrate the decrease of sound by reflection. What are acoustic clouds?*

---

* "The influence of wind on the intensity of sound seems due to the fact that, owing to obstructions opposed by the ground, there is a considerable difference between the velocity of the wind close to the ground and the velocity at the height of a few feet above the ground. Thus in a meadow the velocity of the wind at one foot above the surface may be only half what it is at eight feet above the surface. Let us take the velocity of sound at 1,100 feet per second, and suppose that the velocity of a contrary wind is ten feet per second at the surface, and twenty feet per second at the height of eight feet above the surface. Thus, considering this circumstance alone, the wave of sound at the end of a second would be at the surface ten feet in advance of its position at eight feet above the surface; so that the front of the wave, instead of being a vertical plane, would be inclined to the horizon. Thus the sound, instead of proceeding horizontally, becomes turned upward. It only remains to add that this tilting of the front of the wave is not delayed until the end of a second, but begins at the origin of the sound and increases gradually. Hence a ray of sound, so to speak, instead of traveling horizontally is curved upward, and thus passes over the head of a person stationed at a distance from the origin. A contrary wind then diminishes the intensity of sound by lifting the sound off the ground, and the amount of this lifting increases as the distance from the origin increases. The various consequences which may be deduced from the preceding theory have been verified by experiments.

What is the difference between noise and music? Upon what does pitch depend? Describe the siren. How is it used

Thus it follows that a listener when the wind is contrary may expect to recover a sound, which he has lost at a certain distance from its origin, by ascending to some height above the surface. Also the influence of a wind will be but small if the surface be very smooth; thus sounds are heard against the wind much farther over calm water than over land. Again, suppose the origin of the sound to be elevated above the surface: then if the listener be also raised above the surface he may hear a very loud sound made up of two parts, namely, that which has traveled horizontally, and that which has been tilted upward from the ground by the action of the contrary wind. Next, suppose the wind to be *favorable* instead of *contrary*. In this case the higher part of the wave of sound moves more rapidly than the lower, and so the plane front of the wave is tilted *forward*, and the rays of sound are bent *downward* to the advantage of the listener on the ground. Then the influence of the wind on sound has been shown to depend on the circumstance that when the wind is blowing, the velocity of sound is different at different heights above the ground; similar effects will therefore follow if this difference of velocity is produced by any other cause instead of by the wind. Now change of temperature affects the velocity of sound: if the temperature rise one degree of Fahrenheit's thermometer, the velocity increases by about a foot per second. In general, as we ascend in the air during the day the temperature decreases, and therefore so also does the velocity of sound. Thus the result is the same as in the case of a *contrary* wind; the ray of sound is lifted over the head of a person on the ground, so that the audibility of the sound is diminished. The presence of vapor in the atmosphere also affects the propagation of sound; the velocity increases as the quantity of vapor increases. The direct effect, however, is very slight, but indirectly the vapor is of consequence, for it gives to the air a greater power of radiating and absorbing heat, and so promotes inequality of temperature. The variation of temperature is greatest when the sun is shining, so that it is greater by day than by night, and greater in summer than in winter. Hence, according to the theory now explained, sounds ought to be heard more plainly by night than by day, and more plainly in winter than in summer. That sounds are heard more plainly by night than by day is a well-known fact. We have supposed that the temperature *decreases* as we ascend in the atmosphere; but it may happen on some occasion that the temperature at the surface is *lower* than it is a little above the surface. This may be the case, for instance, over the surface of the sea in the day-time, and over the surface of the land by night. Thus the effect on sound will be similar to that of a *favorable* wind. It is obvious that by the combined influence of wind and temperature the results produced may vary much as to degree; for instance, the operation of a contrary wind may be neutralized by that of the temperature rising as we ascend above the surface." See "Proceedings of the Royal Society of Great Britain," volumes XXII. and XXIV.

to determine the number of vibrations in a sound? How is the octave of any note produced? How can we ascertain the length of the wave in sound? What length of wave produces the low tones in music? The high tones? Give the illustration of the locomotive whistle. When are two tones in unison? How can we find the length of the wave in *any* musical sound? What is meant by the super-position of sound-waves?

How can two sounds produce silence? What is this effect termed? Illustrate interference by means of a tuning-fork. What are "beats"? Describe the vibration of a cord.

Describe the sonometer. What is the object of the wooden box? Give the three laws of the vibration of cords. What is a node? Describe the experiments illustrating the formation of nodes. What are acoustic figures? Nodal lines?

What is the fundamental tone of a cord? A harmonic? What causes the difference in the sound of various instruments? Does a bell vibrate in nodes? The violin-case? A piano sounding-board? State the fractions representing the relative rates of vibration of the different notes of the scale. How is the sound produced in wind-instruments? How is the sound-wave started in an organ-pipe? In a flute? What determines the pitch? What are sympathetic vibrations? Describe the resonance globe. What is a sensitive flame?

A singing flame? Describe the phonograph. The ear. What is the office of the Eustachian tube? Is there any opening between the external and internal ear? What effect does it have on the hearing to increase or diminish the pressure of the air? How does a concussion sometimes cause temporary deafness? How can this be remedied? What are the limits of hearing? Does the range vary in different persons? What sounds are generally heard most acutely? Are there probably sounds in Nature we never hear? What causes the "whispering of the pines"?

VII. Optics.—Define. A luminous body. A non-luminous body. A medium. A transparent body. A translucent body. An opaque body. A ray of light. Show that neither air nor water is perfectly transparent. Why is the sun's light fainter at sunset than at midday? Define the visual angle. Show how distance and size are intimately related.

State the laws of light. Do they resemble those of sound? What is the velocity of light? How is this proved? Explain the undulatory theory of light.

How does light-motion differ from sound-motion? What is diffused light? Why are some objects brilliant and others dull?

Why can we see a rough surface at any angle, and an image in the mirror at only a particular one? Would a perfectly smooth mirror be visible? How does reflection vary? Define mirrors. Name and define the three kinds.

What is the general principle of mirrors? Why is an image in a plane mirror symmetrical? Why is it reversed right and left? Why is it as far behind the mirror as the object is before it?

Why can we often see in a mirror several images of an object? Why can we see these best if we look into the mirror very obliquely? Why is an image seen in water inverted? When the moon is near the meridian, why can we see the image in the water at only one spot? When do we see a tremulous line of light? What is the action of a concave mirror on rays of light? Define the focus. Center of curvature. Focal distance. Describe the image seen in a concave mirror. What are conjugate foci? Describe the image seen in a convex mirror. Why is it smaller than life? Why can it not be inverted like one seen in a concave mirror?

Define Refraction. Does the partial reflection of light as it passes from one medium to another of different density have a parallel in sound? Why is powdered ice opaque while a block of ice is transparent? Give illustrations of refraction.

Why does an object in water appear to be above its true place? What is the general principle of refraction? State the laws of refraction. Explain total reflection. What is the critical angle? Describe the path of a ray through a window-glass. Is the direction of objects changed? Describe the path through a prism.

Name and describe the different kinds of lenses. What is the effect of a double-convex lens on rays of light? What is this kind of lens often called? Describe the image. Why is it inverted after we pass the principal focus? Why is it decreased in size? What is the effect of a double-concave lens on rays of light? Describe the image. Why can it not be inverted like one through a double-convex lens? Describe the images seen in the large vases in the windows of drug-stores. What is Aberration?*

What is mirage? Give its cause.

How is the solar spectrum formed? Name the six principal colors. Show that these six will form white light. Why are the rays separated? What is meant by the dispersive power of a prism? What apparatus possesses this property in a high degree? *Ans.* A triangular bottle filled with a liquid called carbon disulphide ("Popular Chemistry," p. 110). Why does the window of a photographer's dark room sometimes contain yellow glass?

Describe the three kinds of spectra. The spectroscope. What are its uses? Describe rainbows—primary and secondary. Why is the rainbow circular? How is the rainbow formed? Why must it rain and the sun shine at the same time, to produce the bow? Why is the bow in the sky opposite the sun? How many refractions and reflections form the primary bow? The secondary? How many colors can one receive from a single drop? Define complementary colors. How can they be seen? What

---

* To prevent spherical aberration the pupil of the eye can be made very small. The photographer reaches the same result by the use of a diaphragm with a small aperture. "The power of a small orifice to correct the greatest amount of distortion from interfering rays is shown by a simple experiment. The normal eye of an adult can not see to read small print nearer than six inches. Within that distance the type becomes more indistinct the closer it approaches the eye. But if we make a pin-hole through a card and place it close to the eye, we can see to read printed matter of any size even as near as half an inch from the eye. At that distance we can see even the texture of fine cambric with microscopic definition. The cause of this is easily explicable. The rays striking the lens perpendicularly on the center suffer no refraction. The effect of the pin-hole is to exclude all rays but those that impinge perpendicularly on the center of the eye lenses. Hence the image of the object close in front of the eye is pictured on the retina without the interference of the surrounding rays, which would fall obliquely on the lens, and being refracted out of focus would blur the picture. Observation of the effect of a small orifice in correcting aberrant rays, and of the fact that the pupil contracts in near vision, led Haller and some other physiologists to believe that contraction of the pupil was the sole factor in near accommodation. But this view has been sufficiently refuted by other observers."—*Dr. Dudgeon's* "*Human Eye,*" p. 76.

QUESTIONS. 369

is the effect of complementary colors when brought in contrast? (In Fig. 163 opposite colors are complementary.) Why do colors seen by artificial light appear differently than by daylight—as yellow seems white, blue turns to green, etc.

Describe Newton's rings. How are these explained according to the wave theory? What causes the play of color in mother-of-pearl? In soap-bubbles? In the scum on stagnant water? In thin layers of mica or quartz?

What can you say about the length of the waves? State the analogy between color and pitch in music. Why is grass green? When is a body white? Black? What is color-blindness?

What is double refraction? What are the two rays termed? What is polarized light? How does a dot appear through Iceland spar? What other methods are there of polarizing light? State some illustrations and practical uses of polarized light.

What is the meaning of the word microscope? Describe the simple microscope. The compound microscope. How is the power of a microscope indicated? Do we see the object directly in a microscope? Why is the object-lens made so small and so convex?

What is the meaning of the word telescope? Describe the reflecting telescope. The refracting telescope. What is the use of the object-lens? The eye-piece? Is the image inverted? Describe the opera-glass.

The stereoscope. The projecting lantern. How are dissolving views produced?

Describe the Camera. The structure of the eye.* The for-

---

* "In the skate's eye, and generally in the eyes of fishes, the cornea is nearly flat, the aqueous humor is insignificant, and there is virtually no anterior chamber, for the crystalline lens comes up close to the cornea. A convex cornea filled by an aqueous humor would be of no use in the water, the refractive index of the water being identical with that of the aqueous humor. Accordingly, the refraction of the rays of light has to be effected entirely by the crystalline lens, which is nearly spherical, and of much greater refractive power than the corresponding organ in animals which pass their lives in the air. The crystalline lens being so nearly spherical in shape and of such high refractive power, the axis of the eye is short. The eye of the turtle, which is so much in the water, is very similar to that of the fish. The crystalline lens is very near the cornea. The lens is smaller proportionally than that of the skate, nor is it nearly so spherical; and its

mation of an image on the retina. The adjustment of the eye. The cause of near and over sightedness. The remedy. Why do old people hold a book at arm's length? Illustrate the duration of an impression. What is the range of the eye?

VIII. Heat.—Define solar energy. In what ways may it become manifested? What is a diathermanous body? Cold? Gases and vapors? Show the intimate relation between light and heat. What is light? What is the theory of heat? Why can we not see with our fingers or taste with our ears? At what rate does nerve-motion travel? (See "Hygienic Physiology," p. 177.) How long does it take a man, six feet in height, to find out what is going on in his foot?

Name the sources of heat. Describe and illustrate each of these. Can force be destroyed? If apparently lost, what becomes of it? What is Joule's law? Define latent, sensible, and specific heat. Explain the paradox, "that freezing is a warming process and thawing a cooling one." Why does "heat expand and cold contract"? What do you say as to the expansion of solids, liquids, and gases? Illustrate the expansion of solids. Is it better to buy alcohol in summer or in winter? What is the thermometer? Describe it. Describe the process of filling and grading. The F., C., and R. scales. Tell what you can about liquefaction. Of a solid. Of a gas. In one case sensible heat becomes latent, in the other latent heat becomes sensible—why is this?

Explain how a freezing mixture "makes ice-cream." State the theory of vaporization. Of distillation. Since rain comes from the ocean, why is it not salt? Describe the theory of boiling. What is the boiling-point? Do all liquids boil at the same temperature? What would be the effect, if this were the

---

density, and consequently its refractive power, is somewhat less. Hence it has proportionally a longer focus. The cornea is more convex than that of the skate. The fish having no eyelids nor any lachrymal apparatus, its cornea will be apt to become dim by exposure to the air, but the turtle is well supplied with the requisite apparatus for maintaining the transparency of the eye in air. Ophidian reptiles have no eyelids or lachrymal apparatus, but they do not require them, as their cornea is transparent though dry."
—*Dr. Dudgeon's* "*Human Eye*," p. 50.

case? Upon what does the boiling-point depend? Why does pressure raise the melting-point of most substances but lessen that of ice? Why does salt-water boil at a higher temperature than fresh-water? Why will milk boil over so easily? Why will soup keep hot longer than boiling water? Does the air, dissolved in water, have any influence on the boiling-point? Can you measure the height of a mountain by means of a tea-kettle and a thermometer? Show how cold water may be used to make warm water boil. At what temperature will water boil in a vacuum? Why? Can we heat water in the open air above the boiling-point? What becomes of the extra heat? What is the latent heat of water? Upon what principle are buildings heated by steam? Have you ever seen any steam?

Define evaporation. Does snow evaporate in the winter? What can be done to hasten evaporation? Why is a saucepan made broad? Why do we cool ourselves by fanning? Why does an application of spirits to the forehead allay fever? Why does wind hasten the drying of clothes? Describe a vacuum-pan. Why is evaporation hastened in a vacuum? Why is evaporation a cooling process? How is ice manufactured in the tropics? What is the spheroidal state?

Name and define the three modes of communicating heat. Give illustrations showing the relative conducting power of solids, liquids, and gases. What substances are the best conductors? Is water a good conductor? Air? What is the principle of ice-houses? Fire-proof safes? Why do not flannel and marble appear to be of the same temperature? Is ice always of the same temperature? Describe the convective currents in heating water. Where must the heat be applied? Where should ice be applied in order to cool water? Describe the convective currents in heating air. Upon what principle are hot-air furnaces constructed? Ought the ventilator at the top of a room to be opened in winter? At the bottom? Is interplanetary space warmed by the sunbeam?

Does the heat of the sun come in through our windows? Does the heat of our stoves pass out in the same way? Show how the vapor in the air helps to keep the earth warm. Explain the Radiometer. The relation between absorption and reflection.

What is the elastic force of steam at the ordinary pressure of the air? What is the difference between a high-pressure and a low-pressure engine? Which is used for a locomotive? Why? Describe the governor. What is the object of a fly-wheel?

How does the capacity of the air for moisture vary? What is the principle on which dew, rain, etc., depend? Show that a change in density produces a change in temperature. What effect does this have on the temperature of elevated regions? Does an ounce of air on a mountain-top contain the same quantity of heat as the same weight at the foot? How is dew formed?

Upon what objects will it collect most readily? Why will it not form on windy nights? Why is rice-straw used in Bengal in making ice? What is a fog? Why do fogs form over ponds in the early evening? Cause of fogs over the Newfoundland banks? How does a fog differ from a cloud? Why do clouds remain suspended in the air? Describe the different kinds of clouds. Describe the formation of rain. Snow.

How are winds produced? Land-and-sea breezes? Trade-winds? Oceanic currents? Tell about the Gulf Stream. Explain the influence which water has on climate. Of what practical use is the air in water? Describe the apparent exception which exists in the freezing of water. Describe the two processes by which pure water can be obtained. How is an excessive deposit of dew prevented?

IX. Magnetism.—Define Magnetism. A Magnet. A natural magnet. An artificial one. A bar-magnet. A horseshoe magnet. The poles. The magnetic curves. Describe a magnetic needle. What is the law of magnetic attraction and repulsion? Define magnetic induction. Explain it.

When is a body polarized? Give some illustrations of induced magnetism. Does a magnet lose any force by induction? How do you explain the fact that if you break a magnet each part will have its N. and S. poles?

Describe the process of making a magnet. On what principle will you explain this? Describe the compass. Is the needle true to the pole? What causes it to vary? What is the line of no variation? Declination? Why does the needle point N. and S.? What is a dipping-needle? Explain. How is a needle balanced?

Where is the N. magnetic pole? How would one know when he reached it? Does the earth induce magnetism? Which end of an upright bar, in the United States, will be the S. pole?

X.—Electricity. Define frictional electricity. The electroscope. Difference between static and dynamic electricity. Show the existence of two manifestations of electricity. Give the names applied to each.

State the law. What is the theory of electricity? Define a conductor. An insulator.

What is the best conductor? Best insulator? Is a poor conductor a good insulator? When is a body said to be insulated? Can electricity be collected from an iron rod? Describe a plate-glass electrical machine. What is the use of the chain at the negative pole? Define electrical induction. State Faraday's theory.

What is the relation between induction and attraction and repulsion? Describe the electric chime. Explain. Describe the dancing images. The Leyden jar. What gives the color to the spark? How is the jar discharged?

What are the essentials of a Leyden jar? What is the object of the glass? The tin-foil? State the theory of the charging of the jar. Can an insulated jar be charged? Is the electricity on the surface or in the glass? Can the inner molecules of a solid conductor be charged? Will a rod contain any more electricity than a tube? Why is the prime conductor of an electrical machine hollow? What is the effect of points? Describe the electric whirl. Explain the existence of electricity in the atmosphere. What is the cause of lightning? Thunder? Is there any danger after you once hear the report? Describe the different kinds of lightning. Tell how Franklin discovered the identity of lightning and frictional electricity.

Tell what you can about lightning-rods. In what consists the main value of the rod? Does the lightning ever pass upward from the earth? *Ans.* It does, both quietly and by sudden discharge. Has nature provided any lightning-rods? What is St. Elmo's fire? What is the velocity of electricity? Illustrate its instantaneousness. Explain the action of the Voss electrical machine.

Name some of the effects of frictional electricity—(1) Physical, (2) Chemical, (3) Physiological. How are voltaic electricity and chemistry related? Why is voltaic electricity thus named? Tell the story of Galvani's discovery. What was his theory? Give an account of Volta's discovery. How can we form a simple pile? Describe the simple voltaic circuit.

Define the poles. Electrodes. Closing and breaking the circuit. What is necessary to form a voltaic pair? Describe the chemical change. Why does the hydrogen come off from the copper? Tell what you can about the current.

What really passes along the wire? How is this force transmitted? Will a tube, then, *convey* as much electricity as a rod? Explain the term electric potential.

Describe Smee's battery. Grove's battery. The chemical change in this battery. What are the advantages of Grove's battery? Describe Bunsen's battery. Daniell's battery. The Potassium Bichromate battery. Compare frictional and voltaic electricity.

State the effects of voltaic electricity, (1) Physical—heat and light; (2) Chemical—decomposition of water, electrolysis, electrotyping, electro-plating, etc.; (3) Physiological.

What is the effect of a voltaic current on a magnetic needle? What is a galvanometer? An electro-magnet? Show how a coil can be magnetized. How are bar-magnets made? Describe the magnetic telegraph. How is a message sent? How is one received? What is a sounder? What is the general principle of the telegraph? Describe the relay. Name the use of each instrument. Describe a magneto-electric machine. Describe Wilde's machine. What are induced currents? Describe the Telephone. The Microphone. What is the difference between the acoustic and the magnetic telephone? Explain Ruhmkorff's coil. Thermal electricity. A thermo-electric pile. Describe the **electric fish.**

INDEX.

The index figures denote the page.

A

Aberration, 208, 219.
Achromatic, 220.
Acoustics, 153.
Acoustic clouds, 165.
" figures, 174.
Action and reaction, 25.
Adhesion, 49.
Air, 129.
Air-pump, 129.
Alcoholmeter, 116.
Amplitude, 68.
Analyzer, 224.
Annealing, 47.
Archimedes, 97, 135, 149, 239.
" Law of, 115, 149.
Aristotle, 15, 98, 277.
Artesian wells, 111.
Atmosphere, 136, 145.
Atomic theory, 3.
Atomizer, 143.
Attraction, 41.
" of adhesion, 49.
" of cohesion, 43.
" Capillary, 49.
" Gravitation, 55.
Avogadro's law, 16.

B

Bacon, 15, 277.
Barker's mill, 125.
Barometer, 138.
Battery, Bunsen's, 320.
" Grove's, 320.
" Daniell's, 319.

Battery, Potassium Bichromate, 319.
" Thermo-electric, 346.
Beats, 171, 186.
Bell, 156, 175.
Boiling, 252.
Bolometer, 347.
Brittleness, 14.
Britannia Bridge, 105.

C

Caloric, 277.
Camera, 230.
Capillarity, 49.
Capstan, 87.
Cartesian diver, 131.
Caustic, 199.
Center of gravity, 57.
" " oscillation, 69.
" " percussion, 70.
Centrifugal force, 31.
Chemical affinity, 43.
" change, 4.
Chromatic aberration, 219.
Clepsydra, 78.
Clock, 71.
Clouds, 266.
Cohesion, 43.
Coils, Induction, 337.
Color, 216.
" blindness, 216.
" Complementary, 216.
Columns of air, 177.
Compass, 287.
Compensation pendulum, 248.
Condenser, 130.

Conductors, 299.
Conservation of energy, 35.
Cords, Vibration of, 171.
Correlation of forces, 278.
Co-vibration, 158, 179.
Crystals, 46.
Cumulative contrivances, 93.
Current, Electric, 315.
" of rivers, 123.
Curves, Magnetic, 285.

D

Declination, 285.
Democritus, 16, 277.
Dew, 265.
Diathermancy, 275.
Dichroic, 217.
Diffraction, 221.
Diffusion of liquids, 52.
" " gases, 52.
Dissolving views, 229.
Distillation, 251.
Divisibility, 8.
Double refraction, 222.
Ductility, 10.
Dynamo-electric machine, 343.

E

Ear, The, 181.
Ear of Dionysius, 185.
Ear-trumpet, 161.
Echoes, 164.
Elasticity, 12.
Electric battery, 318.
" chime, 304.
" light, 345.
" potential, 297.
" telegraph, 330.
" whirl, 300.
Electrical machine, Plate, 302.
" " Voss, 308.
Electricity, 292.
" Animal, 347.
" Frictional, 294.
" Voltaic, 313.
Electrophorus, 296.
Electro-gilding, 325.

Electro-magnets, 330.
" negative and positive substances, 323.
" plating, 325.
Electrolysis, 322.
Electromotive force, 316.
Electroscope, 294.
Energy, 4, 34.
" Kinetic, 35, 65.
" Potential, 35.
" Radiant, 243.
" Solar, 212.
Equilibrium, 65.
Eustachian tube, 182.
Evaporation, 254.
Expansion, 247.
Extraordinary ray, 223,
Extension, 6.
Eye, The, 230.

F

Falling bodies, 58.
Faraday, 40, 279, 300.
Fire-engine, 140.
Fish, 119.
Flames, Sensitive, 179.
" Singing, 180.
Floating bodies, 118.
Fly-wheel, 93.
Focus, 198, 205.
Fogs, 265.
Force, 21.
" pump, 140.
" Centrifugal, 31.
" Centripetal, 31.
" Molecular, 43.
Forces, Parallelogram of, 27.
" Triangle of, 28.
" Polygon of, 28.
" Composition of, 27.
" Resolution of, 28.
Foucault, 72.
Fountains, 110.
Franklin, 310.
Fraunhofer's lines, 214.
Freezing mixture, 250.
" of water, 271.

INDEX. 377

Friction, 22.
Frost, 265.
Fulcrum, 82.

G

Galvanometer, 329.
Galileo, 40, 77, 98, 149, 187.
Gases, 44.
" Adhesion of, 52, 144.
" Buoyancy of, 135.
" Compressibility of, 13, 137.
" Diffusion of, 52.
" Elasticity of, 13, 137.
" Osmose of, 53.
" Pressure of, 133.
Geissler's tubes, 339.
Gold-leaf, Making of, 11.
Governor, The, 262.
Gravitation, 55.
Gravity, 56.
" Center of, 57.
" Specific, 113.
Guericke, 134, 138.
Gulf Stream, 270.

H

Halos, 219.
Hardness, 14.
Harmonics, 175.
Hay-scales, 86.
Heat, 243.
" affected by rarefaction, 264.
" Absorption of, 260.
" Conduction of, 257.
" Convection of, 258.
" Expansion by, 247.
" Latent, 250.
" Mechanical equivalent of, 246.
" Physical effects of, 247.
" Radiation of, 258.
" Reflection of, 260.
" Specific, 256.
" Theory of, 244.
" unit, 249.
Heating by steam, 276.
Helix, 328.
Helmholtz, 188.

Hiero's fountain, 133.
Holtz's machine, 308.
Horse-power, A, 95.
Huygens, 40, 70, 240.
Hydrodynamics, 121.
Hydraulic ram, 142.
Hydrometer, 116.
Hydrostatics, 101.
Hydrostatic bellows, 108.
" paradox, 109.
" press, 104.

I

Ice-crystals, 46, 268.
Iceland spar, 222.
Imbibition, 50.
Impenetrability, 7.
Inclined plane, 88.
Indestructibility, 10.
Index of refraction, 204.
Induction, 212, 223.
Inertia, 23.
Insulators, 219.
Interference, 128, 169.]
Isochronous, 68.

J

Joule's law, 246.

K

Kaleidoscope, 197.
Kite, 30.

L

Lantern, 228.
Le Conte, 278.
Lenses, 205.
Land-and-sea breeze, 269.
Lever, 82.
Leyden jar, 225.
Light, 191.
" Composition of, 210.
" Diffraction of, 221.
" Interference of, 220.
" Laws of, 192.
" Polarization of, 221.
" Reflection of, 194.

Light, Refraction of, 202.
" Theory of, 193.
" Total reflection of, 208.
" Velocity of, 192.
" Waves of, 193.
Lightning, 310.
Lines of Force, 285.
Liquids, Buoyancy of 118.
" Cohesion of, 43, 44.
" Compressibility of, 13.
" Diffusion of, 52.
" Elasticity of, 13.
" Osmose of, 53.
" Pressure of, 105.
" Specific gravity of, 116.
" Surface tension of, 45.
" tend to spheres, 45.
Liquefaction, 250.
" of gases, 276.
Lissajous, 188.
Locke, 277.

M

Machinery, 81.
Magdeburg hemispheres, 133.
Magnetic curves, 285.
Magnetism, 281.
Magneto-electric machine, 342.
Magneto-induction, 339.
Magnets, 281.
Malleability, 11.
Mariotte's law, 148.
Mass, 14, 23.
Measure, Standards of, 16, 114.
Mechanical powers, 81.
Mechanics, Principle of, 81.
Meteorology, 264.
Meter, 6.
Metric system, 17.
Microphone, 341.
Microscopes, 225.
Mirage, 209.
Mirrors, 194.
Molecules, 3.
Molecular forces, 43.
Moment of a force, 83.
Momentum, 23.

Motion, 21.
" Circular, 30.
" Communication of, 21.
" Composition of, 27.
" Laws of, 22.
" Perpetual, 94.
" Reflection of, 33.
" Resistance to, 22.
Multiple images, 196.
Music, 165.
Musical scale, 176.

N

Near-sightedness, 232.
Needle, Magnetic, 285.
" Dipping, 288.
Newton, 40, 77, 239.
Newton's rings, 220.
Nicol's prism, 224.
Nodal lines, 174.
Nodes, 173.
Noise, 165.

O

Ocean currents, 269.
Octave, 168.
Opera-glass, 228.
Optics, 191.
Optical instruments, 225.
Ordinary ray, 223.
Organ pipes, 178.
Oscillation, Center of, 69.
Osmose of gases, 53.
" " liquids, 53.
Oversightedness, 232.
Overtones, 175.

P

Pascal, 103, 150.
Pendulum, 68.
Percussion, Center of, 71.
Perpetual motion, 94.
Phonograph, 180, 188.
Pinion, 88.
Pisa, Tower of, 67.
Pitch, 166.
Platinum wire, 11.

INDEX. 379

Plato, 15.
Plumb-line, 56.
Pneumatics, 129.
Pneumatic inkstand, 142.
Polariscope, 223.
Polarization of light, 221.
" Electric, 301, 318.
" Magnetic, 284.
Polarizing angle, 223.
Porosity, 8.
Pressure of air, 133.
Prince Rupert's drop, 48.
Prisms, 204.
Projecting lantern, 228.
Pulley, 91.
Pumps, 139.
" Air, 129.
" Force, 140.
" Lifting, 139.
" Sprengel's air, 148.
Pythagoras, 186.

R

Radiometer, 258.
Rain, 267.
Rainbow, 217.
Reaction, 25.
Reaction-wheel, 125.
Reflected motion, 33.
Reflection, Total, 208.
Refraction, Index of, 204.
Relay, 334.
Resonance, 179.
Rivers, 123.
Ruhmkorff's coil, 337.
Rumford, Count, 278.
Rupert's drop, 48.

S

St. Elmo's fire, 311.
Screw, 90.
Sensitive flames, 179.
Ship, Sailing of, 29.
Singing flames, 179.
Siphon, 141.
Siren, 166.
Size, 14.

Snow, 267.
Solution, 51.
Sonometer, 172.
Sound, 153.
" Intensity of, 160.
" in a vacuum, 156.
" Interference of, 169.
" Loudness of, 160.
" Production of, 153.
" Reflection of, 162.
" Refraction of, 161.
" Transmission of, 154.
" Velocity of, 158.
Sounding-boards, 172.
Sound-waves, 154.
Speaking-tubes, 161.
" trumpet, 161.
Specific gravity, 113.
" " flask, 116.
Spectroscope, 213.
Spectrum, Prismatic, 210.
" Normal, 212.
" Interruptions in, 213.
" Kinds of, 214.
" Analysis, 214.
Spherical aberration, 208.
Spheroidal state, 256.
Steam, 252.
" engine, 261.
Steelyard, 85.
Stereoscope, 234.
Stringed instruments, 172.
Surface tension, 45.

T

Tacking, 30.
Tackle-block, 93.
Telegraph, 330.
Telephone, 340.
Telescope, 226.
Temperature, 248.
Tempering, 47.
Tenacity, 12.
Thermo-electricity, 346.
Thermometers, 248.
Thunder, 310.
Torricelli, 135, 149.

380 INDEX.

Torsion pendulum, 13.
Total reflection, 208.
Tourmaline, 222.
Trade-wind, 269.
Tubes, 122.
Turbine wheel, 124.
Tyndall, 188.

V

Vaporization, 251.
Velocity, 21.
Velocity of heat, 243.
" " light, 192.
" " sound, 158.
Vibration, 68.
Vibrations of air, 154.
" " cords, 171.
" " ether, 193.
" " pendulum, 68.
" Sympathetic, 179.
Virtual velocity, 98.
Vision, 232.
" Binocular, 233.
Visual angle, 191.
Vocal Memnon, 185.
Voltaic arc, 321.
" battery, 318.

Voltaic electricity, 313.
" pair, The, 314.
Volume, 6.
Voss' machine, 308.

W

Watches, 78.
Water, 270.
" barometer, 138.
" level, 112.
" wheels, 123.
Waves, 126, 154.
Wave motion, 126.
Wedge, 91.
Weight, 57, 58.
Welding, 44.
Wells, 111.
Wheel and axle, 86.
Wheel-work, 88.
Whirligig, 125.
Wilde's machine, 343.
Winds, 268.
Wind instruments, 177.

Y

Young, 240.
Youmans, 278.

www.ingramcontent.com/pod-product-compliance
Lightning Source LLC
Chambersburg PA
CBHW032023220426
43664CB00006B/345